THE MOST SPECTACULAR SKY EVENT OF THE CENTURY

Rare, beautiful, and sometimes frightening, comets make the whole world sit up and take notice. To prepare for the next flash of Halley's Comet through our skies, famous science writer Isaac Asimov brings you the whole story of this fascinating sky visitor.

You'll find out:

- What Halley's Comet is made of
- Where it comes from and where it goes
- Outrageous ideas and bizarre beliefs of ancient peoples and early scientists
- What comet created the biggest meteor shower in modern times
- What terrifying event shook the earth in 1908
- How comets could have killed off the dinosaurs
- And much, much more....

Asimov's Guide to Halley's Comet

"ASIMOV IS ONE OF OUR NATURAL WONDERS AND NATIONAL RESOURCES." –*St. Louis Post-Dispatch*

"[ASIMOV IS] UNDOUBTEDLY THE MOST LUCID SCIENTIFIC POPULARIZER OF HIS DAY." –*Kirkus Reviews*

"MR. ASIMOV IS A PHENOMENON...HE PRACTICES WITH A HIGH GOOD HUMOR AND BRASH PLEASURE THE ROLE OF INFORMAL EXPLAINER-AT-LARGE TO THE UNIVERSE. HE IS ONE OF NATURE'S IRREPRESSIBLE ENTHUSIASTS."

–*The Wall Street Journal*

This spectacular photograph of Halley's Comet and its tail was taken on April 4, 1910.

Asimov's Guide to Halley's Comet

ISAAC ASIMOV

A DELL TRADE PAPERBACK

A DELL TRADE PAPERBACK
Published by
Dell Publishing Co., Inc.
1 Dag Hammarskjold Plaza
New York, New York 10017

I wish to acknowledge the assistance of Sandra Kitt for picture research and Thomas A. Lesser, Allen Seltzer, and Larry Brown, who assisted her in this work.

For list of illustrations and credits, see pages 111-113.

Copyright © 1985 by Nightfall, Inc.

All rights reserved. No part of this book may be reproduced or transmitted in any form or by any means, electronic or mechanical, including photocopying, recording, or by any information storage and retrieval system, without permission in writing from the Publisher. For information address Walker and Company, New York, New York.

Dell ® TM 681510, Dell Publishing Co., Inc.

ISBN: 0-440-50434-1

Reprinted by arrangement with Walker and Company
Printed in the United States of America

October 1985
10 9 8 7 6 5 4 3
MV

To Richard Winslow,
who suggested this book so forcefully

Contents

1 The Fear of Comets 1
2 The Paths of Comets 8
3 The Return of Halley's Comet 22
4 Dim Comets 36
5 The Death of Comets 45
6 Nineteenth-Century Comets 55
7 Comet Tails and Meteors 64
8 The Distant Comets 79
9 The Birth of Comets and the Solar System 87
10 Comets and Catastrophes 98
List of Illustrations 111
Index 115

Asimov's Guide to Halley's Comet

This nineteenth-century drawing suggests how the ancients often considered comets as symbols of giant swords.

1

The Fear of Comets

Comets have always seemed to frighten people.

For one thing, they don't seem to follow the rules. The other heavenly bodies—the stars, the Sun, the Moon, the planets—all move in regular fashion.

The stars are particularly regular, all moving round and round the sky with a steady, even motion that does not change. While they do so, their positions relative to each other do not change, either.

The Sun and the Moon are not quite so even. The noonday Sun drifts slowly higher and lower in the sky as the year progresses, and the Moon changes its shape from night to night. The planets change their speeds and even their direction of motion as time passes. These changes, however, are regular. They can be worked out, and the position of these heavenly bodies can be predicted far into the future.

Not so comets. A dim comet will appear in the sky without warning. It will get brighter and brighter for a while, then it will grow dimmer and dimmer and will disappear. Another comet may not appear for fifty years or more, or one may appear the very next year. The ancient astronomers, who worked out the motion of all the other objects in the sky, were helpless where comets

Two early astronomers observe the Comet of 1596 as depicted on the title page of a German pamphlet.

were concerned. They couldn't tell when one would appear, or in what part of the sky it would appear, or how long it would remain visible.

This was important because the ancients believed it was possible to foretell the future from the positions of the Sun, the Moon, and the various planets against the background of the stars. The pattern of these positions changed from night to night and year to year, and it was thought that this represented a code that wise men could interpret and work out for the guidance of human beings. (This belief, which is quite wrong, by the way, is called "astrology," and it still influences many superstitious people.)

Comets, coming along unexpectedly, seemed to be a warning of something unusual. To most people "something unusual" means disaster, so the sight of a comet was frightening.

The fearful Comet of 1528 as seen by a French physician.

This was reinforced by the shape of comets. The Sun is a bright circle. The Moon has a variety of shapes, but at least half the edge is the arc of a circle. All the other bodies are points of light. A comet, however, is a luminous circle of haze and from one side stretches outward a dim, slightly curved line of haze, like a long tail or streaming hair. The Greek word for "hairy" is "kometes," and that is where our word "comet" comes from.

One sign of mourning in ancient times was for women to loosen their hair and allow it to stream down their backs, as an indication that they were too upset to take care of it. It was easy to see a comet, then, as a shrieking, mourning woman with her hair flying in the wind. What else could that be but a sign of disaster to come?

Once people were sure that comets were signs of disaster, they imagined the tail to be in the shape of a sword or saber, and the circle of haze to be a decapitated head. Writers competed with each other in inventing gruesome descriptions, so the fear of comets increased.

When a comet appeared, people would record the year of its appearance and would then describe the terrible events that took place soon afterward. This was considered "proof" that comets were omens of disaster. Of course, terrible events take place every year, whether comets appear or not, so that wasn't proof of anything.

For instance, a comet appeared in the sky in 44 B.C. and that was later taken to be connected with the assassination of Julius Caesar in that very same year. Again, a comet in 11 B.C. was supposed to have had something to do with the death of Marcus Agrippa, a Roman statesman, the year before. A comet in A.D. 837 was supposed to warn of the death of the Emperor Ludwig the Pious three years later.

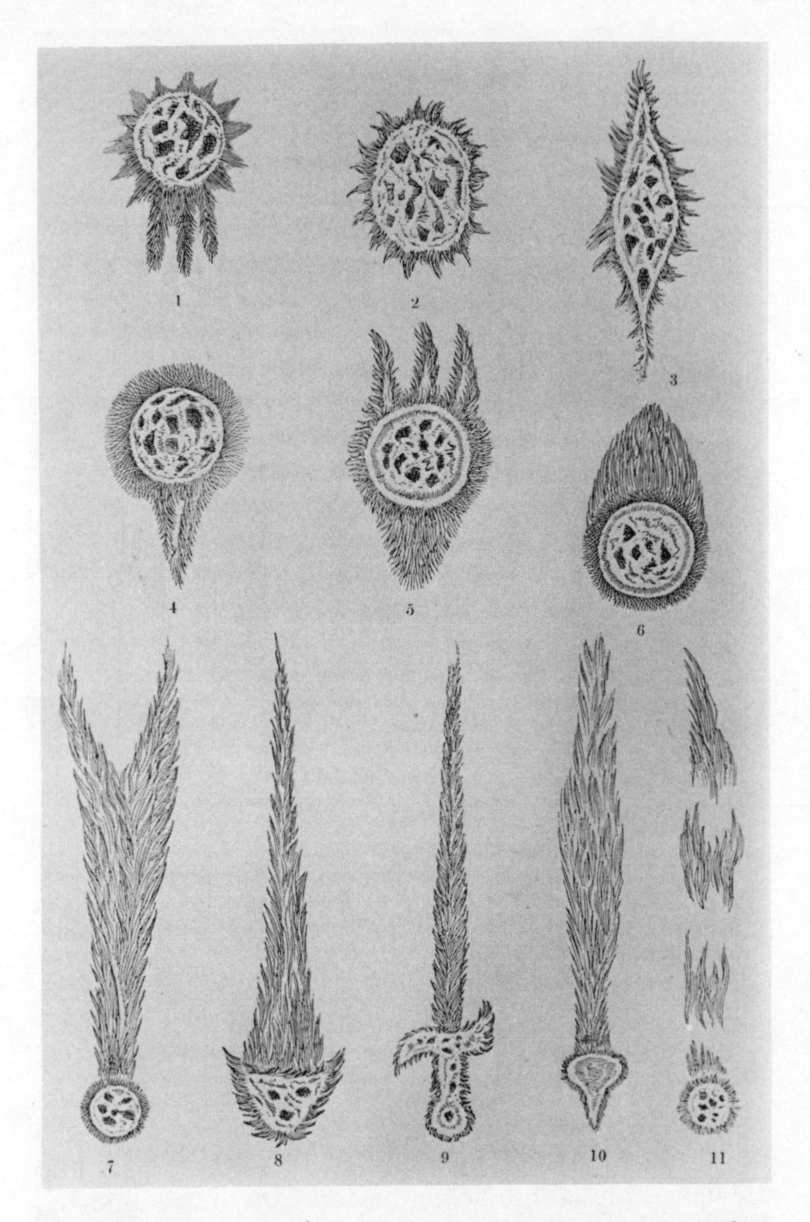

Typical of everyone in ancient times, the Roman writer Pliny described comets in terms of supernatural weapons—stones, discs, swords, and daggers. This drawing was made by Hevelius, a seventeenth-century German astronomer.

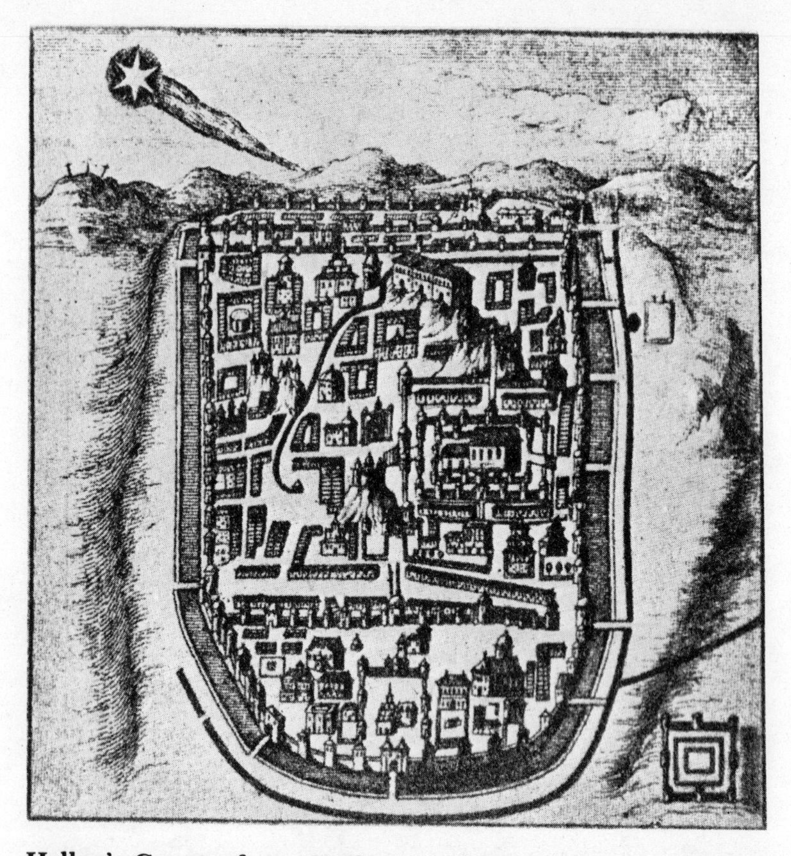

Halley's Comet of A.D. 66 shown over Jerusalem in this seventeenth-century print. The Comet was regarded as an omen predicting the fall of the city to the Romans which actually occurred four years later.

Nor did they always foretell the death of rulers. Sometimes comets were associated with war. A comet seen in A.D. 66 was later said to be a warning of the fall of Jerusalem to the Romans in A.D. 70. Another comet, in 1066, was thought to be connected with the conquest of England by William of Normandy later that year. Still another, in 1456, was viewed as a heavenly comment on the fall of Constantinople to the Turks in 1453.

As you can well imagine, any comet at any time could be seen as signifying disaster if you searched history books for several years before and after. Nor did people seem to notice that comets meant good news as well as bad. The conquest of England was good news to the Normans, and the fall of Constantinople was good news to the Turks.

Nearly 900 comets were reported during the centuries before the telescope was invented, and they were unusually abundant between 1400 and 1600. In those two centuries, about twenty very bright comets were seen and, as it happened, this was a particularly disturbed time. The Turks had taken over all of southeastern Europe and at one point had penetrated as far as Vienna and laid siege to it. The Protestant Reformation split the Church in two, and a cycle of vicious "wars of religion" began. All those comets must have seemed very significant to the people of western Europe.

There were, of course, a few people who argued against this comet fear, maintaining that comets had nothing to do with events on Earth and were merely natural phenomena. Such views exerted no influence, however. Every new comet set off a new rash of books describing all the horrors that comets foretold, and these were much more popular because they were more dramatic and harrowing.

The result was that, throughout ancient and medieval times, no Europeans made any really scientific observation of comets. The only reaction was one of terror and expectation of disaster.

2

The Paths of Comets

Finally, in 1472, an astronomer treated a comet as an astronomic phenomenon and viewed it calmly. The astronomer was a German named Johann Müller (1436–76), who is better known as Regiomontanus (the Latin translation of Königsberg, the city of his birth). He and a student of his observed the comet from time to time and noted its position against the stars. This enabled them to draw an imaginary line across the sky, one that marked the comet's path. (It seems strange that no one had ever thought to do such a thing before.)

By the time a comet appeared in 1531 (with two more comets in 1532, and one each in 1533, 1538, and 1539), there were additional astronomers who managed to observe calmly. One was an Italian, Girolamo Fracastoro (1483–1553). He observed the Comet of 1531 and the four that followed, and in 1538 published a book in which he stated that the tails of comets always pointed away from the Sun.

A German astronomer, Peter Apian (1501–52), also studying these comets and not knowing of Fracastoro's

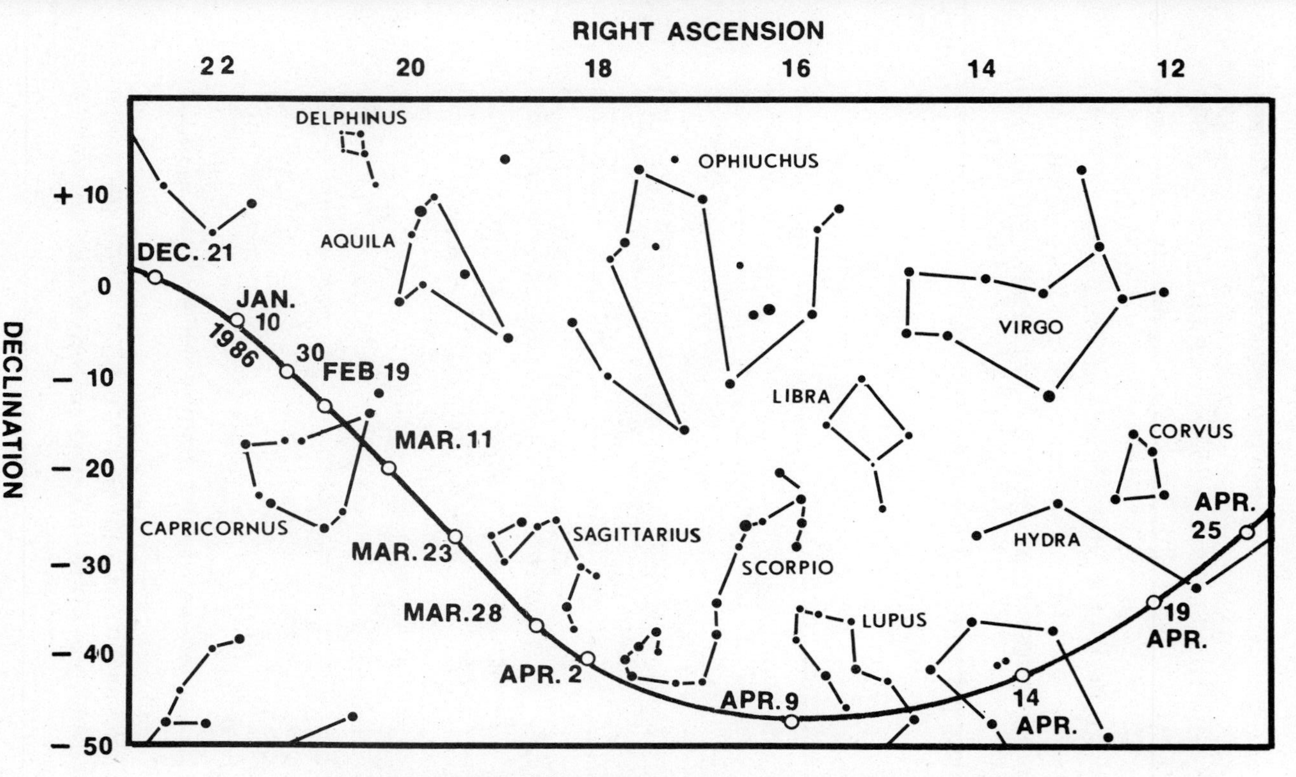

Halley's 1985–86 path through the constellations as seen by an observer looking to the South. Viewers north of New York, Chicago, and San Francisco will not be able to see the comet south of −40° declination.

work, published a book in 1540 in which he came to the same conclusion. In his book Apian published the first scientific drawing of a comet, one in which he indicated the position of the tail with reference to the Sun.

For the first time, comets were made to seem not completely lawless. At least the tail always followed a simple rule. Comets began to seem just a bit more like ordinary astronomical objects.

But were they?

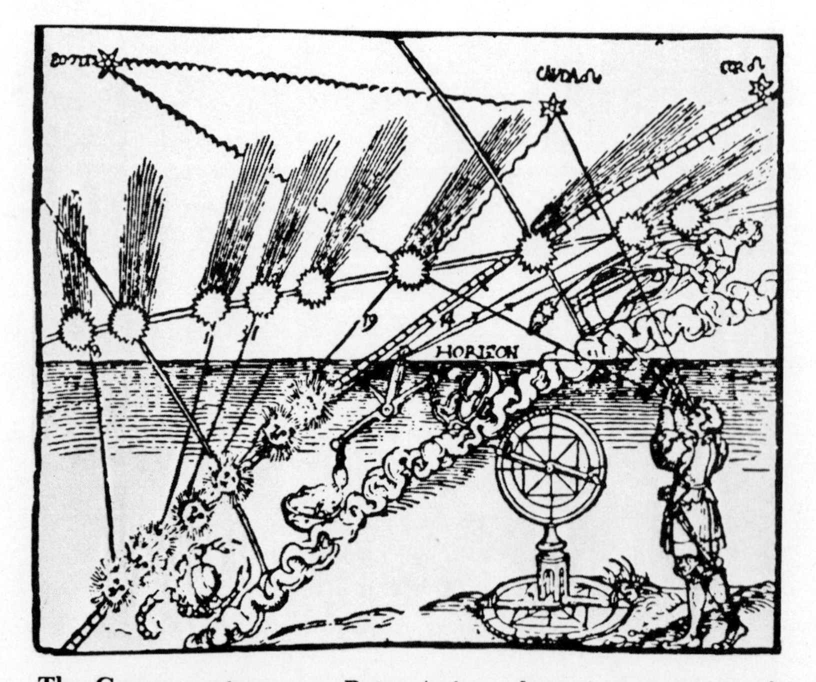

The German astronomer Peter Apian, observing positions of the Comet of 1531, was the first to publish a scientific drawing of comet tails always pointing away from the Sun.

The Greek philosopher Aristotle (384–322 B.C.) hadn't thought so. He believed that all the heavenly bodies moved in stately, predictable courses round and round the Earth. Since comets were unpredictable, coming and going in totally unexpected fashion, he reasoned that they could not be among the heavenly bodies. Instead, he thought, they must be slowly burning flames in the upper atmosphere. These would kindle now and then, burn for some weeks or months, then slowly be extinguished.

Since comets, according to this view, were part of the atmosphere, they must be closer to the Earth than the heavenly bodies that lay beyond the atmosphere. They must be closer to the Earth than even the Moon, which the Greeks considered (correctly) to be the closest to Earth of all the heavenly bodies.

Such was Aristotle's prestige that his views were accepted for nearly two thousand years after his death.

But then, in 1577, came another comet, and this was observed by a Danish astronomer, Tycho Brahe (1546–1601). It occurred to Tycho (who is usually known by his first name) that he might be able to determine the distance of the comet.

The distance of objects in the sky can be estimated by the principle that nearby objects seem to shift position against a distant background if viewed from different places. Thus, if you hold your finger in front of your face and close your right eye, you will see your finger in a certain position against objects in the background. If you keep your finger in the same place but open your right eye and close your left, the finger will seem to shift position against the background.

The further you hold your finger from your eye, the

smaller the distance through which it will shift. From the size of the shift (or "parallax"), then, distance can be calculated.

Thus, the Moon will seem to shift position against the much more distant stars if it is viewed from different places (say, Rome and London). Allowing for the distance between the places from which it is viewed, the Moon's parallax can be used to calculate its distance.

This was what Tycho tried to do in connection with the Comet of 1577. He recorded its positions against the stars night after night and compared these with the positions reported by astronomers in other places. Tycho found no parallax that he could detect. This meant to him that the comet had to be at least four times as far away as the Moon. If it were closer than that, there would have been a parallax large enough to detect.

After that, there was no question but that comets were astronomical bodies, just as the planets were.

In 1543, meanwhile, a Polish astronomer, Nicolaus Copernicus (1473–1543), had published a book in which he reasoned that the Sun and the planets did not circle the Earth as Aristotle and other Greek philosophers had believed. Instead, Copernicus said, it was easier to understand the way in which the planets moved across the sky if they were viewed as circling the Sun. The Earth itself was a planet and it too, along with its accompanying Moon, circled the Sun. Hence we now talk of the "solar system," from the Latin word "sol" meaning "sun."

Then, in 1609, a German astronomer, Johannes Kepler (1571–1630), who had worked as assistant to Tycho during the latter's last years, showed that the planets, including Earth, circled the Sun in paths (or "orbits")

The Great Comet of 1577. The study of this comet helped Tycho Brahe establish scientifically that comets are astronomical bodies, not supernatural omens.

that were ellipses, with the Sun at one focus. (An ellipse can be viewed as a kind of flattened circle, with two foci, one on each side of the center. The more flattened the ellipse, the more "eccentric"; that is, the farther apart the foci are, and the nearer they are to the opposite ends of the ellipse.)

The planetary orbits are not very eccentric, so the foci are not very far from the center, and that means the Sun is located not very far from the center.

The question soon arose as to whether comets moved around the Sun as the planets did, and if so in what kind of path. From even casual observations, it certainly didn't seem as though a comet's orbit was anything like a planet's orbit.

Kepler had observed a comet that appeared in 1607, and it seemed to him that it moved in a straight line. He suggested that comets approached from a far distance, moved through the solar system, and then vanished in the far distance in the opposite direction. It appeared once it was close enough to see, grew brighter and brighter till it passed us, then grew dimmer and dimmer, and finally disappeared when it was no longer close enough to see.

In 1609, however, the same year in which Kepler had worked out his scheme of elliptical orbits, the Italian scientist Galileo Galilei (1564–1642), who is usually known by his first name only, constructed a simple telescope and turned it on the heavens. At last human

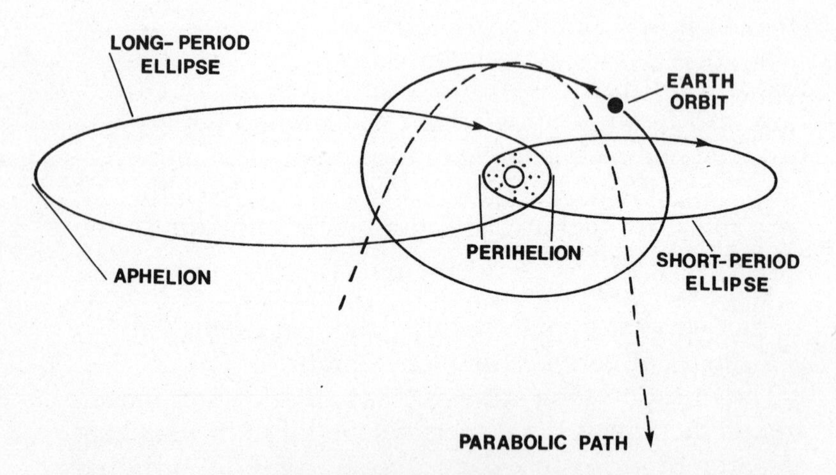

Elliptical and parabolic paths of comets as they pass around the Sun. A comet on an elliptical path will return periodically; one on a parabolic course passes into outer space, never to be seen again. (The Earth's orbit in this perspective appears far more elliptical than it actually is.)

beings could see things in the sky more sharply, more brightly, and in better detail than they could with the eye alone.

Soon every European astronomer had a telescope he could use; the comet that appeared in 1618 was the first to be viewed telescopically. The viewer, a Swiss astronomer, Johann Cysat (1586–1657), agreed with Kepler that comets moved in straight lines.

Not everyone accepted this. An Italian scientist, Giovanni Alfonso Borelli (1608–79), carefully observed the changing position of a comet that appeared in the sky in 1665. It seemed to him that its path across the sky did not make sense if he assumed that the comet was traveling in a straight line.

Borelli felt that a comet might approach the solar system in pretty much a straight line, but that this line curved as the comet neared the Sun. The line of motion veered about the Sun, and then the comet receded in what seemed more and more a straight line. In short, the comet's motion was not like that of an I, but like that of a U, with the Sun located inside and near the bottom of the U. Such a U-shaped orbit is called a "parabola."

Either way, whether its path was an I or a U, any comet would pass through the solar system only once, approaching from infinite distance and receding into infinite distance.

The notion of "infinite distance" was an uncomfortable one, however. It occurred to some scientists that comets might travel in orbits that were very long and very flattened ellipses. The other end of the ellipse would be so far away that the comet would remain unseen in that part of its orbit for many years. Only when it approached the end of the ellipse that was near the

Sun would it be seen. That near end would have a U shape, for one end of a very long ellipse is very much like a parabola in shape.

The difference between a true parabola and a very long ellipse is that a comet following a parabolic orbit never returns, while one in a long elliptical orbit will return eventually, although perhaps not for many, many years. The first to suggest that comets might return periodically, as they moved round long ellipses, was the German scientist Otto von Guericke (1602–86).

The suggestion was interesting, but seemed useless at first. There appeard to be no way in which one might work out the orbit of a comet and *predict* when it would return. Not until cometary motions became as predictable as planetary motions could comets be accepted as true members of the solar system.

In 1687, the year after Guericke died, the English scientist Isaac Newton (1642–1727) published a book in which he established the law of universal gravitation. For the first time astronomers learned the rules that defined, quite accurately, the paths that astronomical bodies *must* take in the neighborhood of other bodies.

By Newton's laws, a comet could travel around the

The complete circuit of Halley's Comet from 1948 to the year 2024 when it will again be 3,250 million miles from the Sun. It was first spotted on its current return, October 20, 1982, by electronic cameras attached to the 200-inch telescope on Mt. Palomar in California. After swinging around the Sun, it will come within 39 million miles of the Earth on its way back to its aphelion.

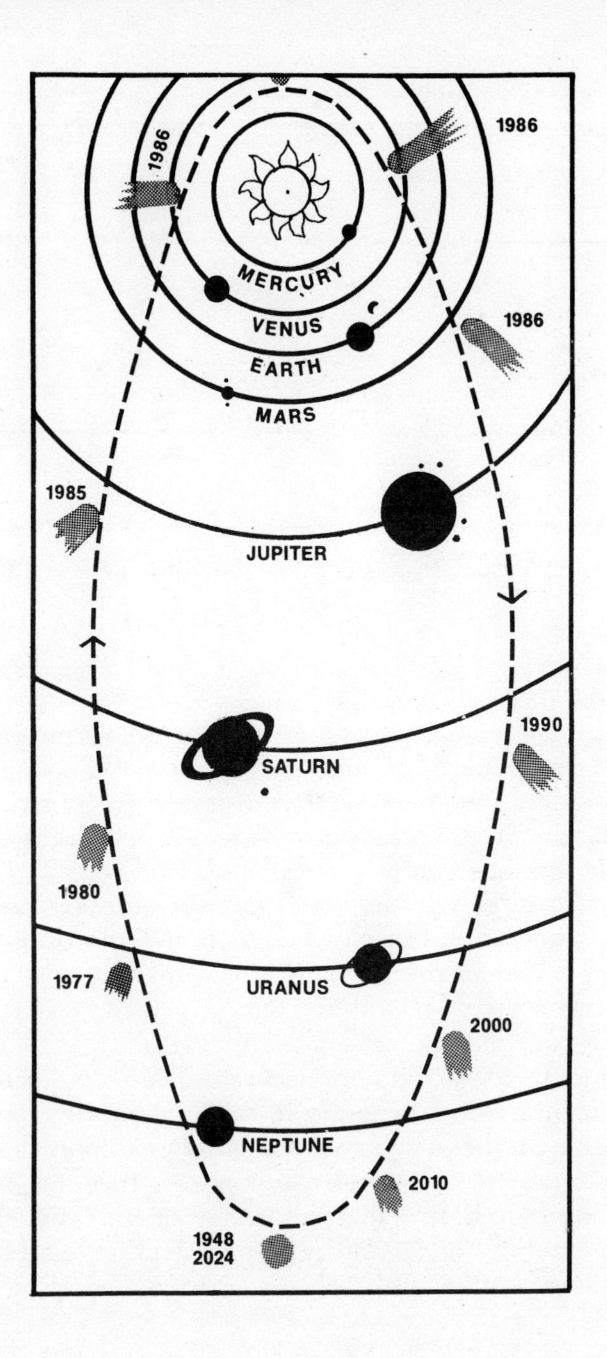
1986
1986
MERCURY
VENUS
1986
EARTH
MARS
1985
JUPITER
1990
SATURN
1980
1977
URANUS
2000
NEPTUNE
2010
1948
2024

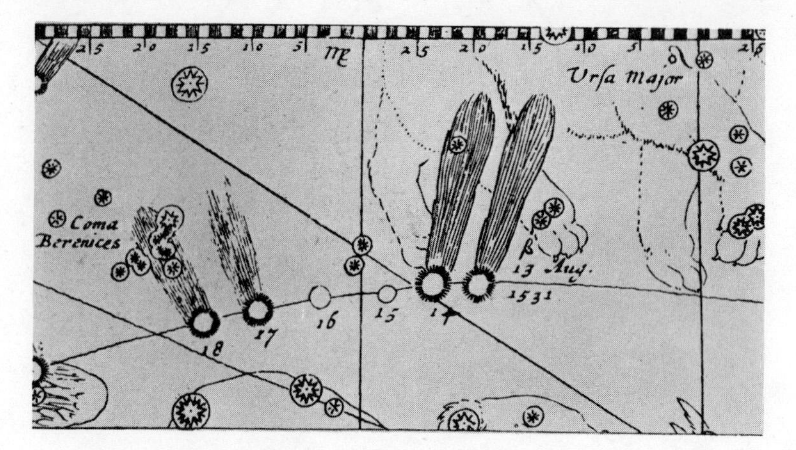

Halley's Comet passed under the Great Bear in 1531.

Sun in a very long ellipse, or in a parabola. One can tell which is which by noting the distance of a comet from the Sun and its speed at that distance. If the motion is slow enough, the orbit *must* be an ellipse and not a parabola, and the comet *must* someday return.

The English astronomer Edmund Halley (1656–1742) was a good friend of Newton's. In 1682 he observed a bright comet in the sky and began his attempts to calculate its orbit. It was not an easy task, and Halley worked at it on and off for many years.

To make his work more accurate, Halley collected all the cometary data he could find, studying the changing positions in the sky of two dozen other comets that had been reported by earlier astronomers. In doing so, he couldn't help noting that the Comet of 1607, which had been carefully observed by Kepler, had traveled across the same portion of the sky as had the Comet of 1682, which Halley had observed. As a matter of fact, the Comet of 1531, which Fracastoro and Apian had observed, had *also* moved across that part of the sky, and so had the Comet of 1456, which Regiomontanus had observed.

To his surprise, Halley noticed that from 1456 to 1531 there was a lapse of 75 years; from 1531 to 1607, one of 76 years; and from 1607 to 1682, one of 75 years. It seemed to Halley that the four comets of 1456, 1531, 1607, and 1682, were actually the same: one that fol-

tel its Return. And, indeed, there are many Things which make me believe, that the Comet which *Apian* obſerv'd in the Year 1531, was the ſame with that which *Kepler* and *Longomontanus* more accurately deſcrib'd in the Year 1607; and which I my ſelf have ſeen return, and obſerv'd in the Year 1682. All the Elements agree, and nothing ſeems to contradict this my Opinion, beſides the Inequality of the Periodic Revolutions. Which Inequality is not ſo great neither, as that it may not be owing to phyſical Cauſes. For the Motion of Saturn is ſo diſturbed by the reſt of the Planet

Excerpt from Halley's first published prediction in which he demonstrated that the Comets of 1531, 1607, and 1682 were one and the same.

Portrait of Edmund Halley (1657–1742) in his later years when he became Britain's Astronomer Royal.

Halley's Comet in 1456 had its head in the Twins constellation and its elongated tail draped across the constellations of the Crab and the Lion. The tail was so long that it stretched a third of the way across the sky.

lowed an orbit that was a long ellipse—so long that the comet came back into view at the near end of the ellipse only every 75 or 76 years.

That was a very daring thought, and Halley hesitated for a long time before making it public. By 1705, however, he had made all the tedious calculations and had convinced himself that it was so. He published his table of various cometary paths and announced his belief that the Comet of 1682 would return in the year 1758.

In a way, it was a frustrating announcement for Halley, for he would have to live to be 102 years old in order to see personally whether his prediction was correct or not. Of course he didn't. He died in 1742, two months past his eighty-fifth birthday.

3

The Return of Halley's Comet

Halley's prediction caused a sensation, but it couldn't remain a sensation. There was, after all, nothing to do but wait for over half a century to see if the comet would indeed return. Very few of the astronomers of the day could count on being alive by then, so they knew they would never know if the prediction was right or wrong. Naturally, they turned to other things.

But 1758 came at last and the months passed, one by one, and no comet appeared in the sky.

It wasn't expected, however, that the comet would necessarily return at the very moment predicted. After all, between the times when the Comets of 1531 and 1607 passed the Sun there had been an interval of 76 years and 1 month, while between the Comets of 1607 and 1682 there had been an interval of 74 years and 11 months. That's a difference of a year and two months. One could not necessarily pin down the next return to 1758. It could be 1759 or even 1760.

But why should there be this irregularity?

If the comet and the Sun were the only two objects involved, then the comet would return with clocklike regularity; but the comet and the Sun are *not* the only objects involved. As the comet moved along its orbit it could pass fairly close to the two large outer planets, Jupiter and Saturn, and these could exert gravitational pulls upon it, speeding it up a bit or slowing it down.

Halley had worked out an orbit for the comet, but there was room for improvement. As the time for the return approached, two French astronomers, Alexis Claude Clairault (1713–65) and Joseph Jérôme Lalande (1732–1807), went over Halley's figures and worked out the orbit of the comet still more carefully. They then allowed for the gravitational pulls of Jupiter and Saturn, as the comet passed them. They calculated that the comet would be a little delayed and would not reach its closest point to the Sun ("perihelion") until April 13, 1759. Naturally, though, it should be spotted in the course of its approach months before perihelion.

Nevertheless, professional astronomers were not sufficiently excited about the prospect to mount an elaborate search for the returning comet.

Astronomy is one science, however, which, to this very day, has its dedicated amateurs who can do important and useful work. One amateur astronomer who was active in 1758 was a well-to-do German farmer named Johann Georg Palitzsch (1723–88). He was familiar with Halley's work and was sure the comet would return.

In November 1758 he set up his telescope and trained it on the part of the sky where the comet ought to appear, if it did return. He waited patiently, and on December 25, 1758, he had what was surely the greatest Christmas of his life, for it was on that day that he became the first person to detect the returning comet.

Palitzsch's report roused the professionals. On January 21, 1759, the first professional sighting was made by a French astronomer, Charles Messier (1730–1817), who was held back for weeks by a siege of bad weather that made viewing very unsatisfactory.

Thereafter, the comet grew steadily brighter, crossed the sky in its appointed path, and remained visible (except when it was very near the Sun) until the end of May. It passed perihelion on March 13, a month earlier than the prediction of Clairault and Lalande.

Why the discrepancy? Well, for one thing, Clairault and Lalande did not know of the existence of the distant planets, Uranus and Neptune, and could not take their gravitational pulls into account. Nor did they have very accurate figures for the masses of Jupiter and Saturn. But considering how much information they lacked, they did very well.

Nowadays, of course, we know all the planets that lie in the comet's path and have excellent figures for their masses. We also have accurate knowledge of the comet's orbit. We know that at its nearest approach to the Sun, the comet is only 55 million miles away, so that it approaches the Sun more closely than Venus does. Thirty-seven years after perihelion, the comet is at "aphelion," which is its farthest distance from the Sun. It is then 3,250 million miles from the Sun. This is over 3.5 times as far away as Saturn, the most distant planet known in Halley's time. Its aphelion is 500 million miles outside the orbit of Neptune, the most distant large planet known today.

Ever since the comet returned in 1758, it has been known as "Halley's Comet." Nowadays it is becoming customary to put the word "comet" first, so that it may also be referred to as "Comet Halley."

The name "Halley" rhymes with "valley," by the way,

at least in American usage. Some years ago, someone named Bill Haley (rhymes with "daily") organized a musical group he called "Haley's Comets." Since then, those people who are more familiar with rock music than with astronomy refer to the comet as "Halley's Comet" with a long "a," but that is wrong and grates on astronomical ears. It is "Halley's Comet," short *a*! (To be sure, there is some feeling in England that Halley pronounced his name "Hawley," but let's ignore that.)

Once Halley's Comet was recognized as a respectable member of the solar system, capable of returning in a periodic and predictable way, astronomers began to look backward to see which of the numerous comet sightings were actually earlier visits of what had now suddenly become the most famous of all comets.

We can start with the perihelion of March 13, 1759, and count backward, through earlier perihelia:

September 15, 1682. This was the comet observed by Halley.

October 27, 1607. This was the comet studied by Kepler, who suggested as a result that comets traveled in straight lines.

August 25, 1531. This was the comet observed by Fracastoro and Apian, who observed that the tail always pointed away from the Sun.

June 9, 1456. This was the comet observed by Regiomontanus. It was a time of great comet fear, too, for Halley's Comet appeared three years after the fall of Constantinople, and it was thought to announce further victories by the Turks. Pope Calixtus III ordered special prayers that might avert God's anger and prevent the Turks from overrunning all of Europe.

November 9, 1378. On this occasion, Halley's comet

was not very bright. This can occur if the Earth happens to be on the opposite side of the Sun from the comet when the latter is making its close approach. The two are then separated by an unusually large distance, and besides, the comet is too close to the Sun to be viewed easily. Sometimes, too, the comet can be so situated that it is not clearly visible in the northern hemisphere, where most of the observers and astronomers are. And sometimes the position of Earth and comet is such that we don't see the tail broadside, and without a good view of the tail the comet is not very spectacular.

October 23, 1301. This was a very bright appearance and was thought to have been observed by the Italian artist, Giotto di Bondone (1267–1337), another of those known to history by his first name. In 1304 Giotto completed a great painting, *The Adoration of the Magi*, in which the wise men are worshipping the newlyborn Jesus. The Star of Bethlehem is usually shown above the manger in such paintings, and Giotto represented it by a realistic picture of a comet, probably drawing it according to his memory of the recent appearance of Halley's Comet. A number of people in those days believed that the Star of Bethlehem had been a comet.

October 1, 1222. Notice that this perihelion is just 79 years before the one above. It turns out on close study that the orbital period of Halley's Comet is just under 77 years, on the average, but thanks to the gravitational pulls of the planet it can vary from 2.5 years less than that to 2.5 years more than that; that is, anywhere from 74.5 to 79.5 years.

April 22, 1145. There is not much to say of this appearance.

March 23, 1066. This is the most famous appearance

Giotto painted the Star of Bethlehem as a comet in his celebrated fresco—"The Adoration of the Magi"—which was finished in Florence three years after Halley's Comet's appearance in 1301.

The artists of the Bayeux Tapestry rendered Halley's Comet of 1066 as a star with a multiple tail. This scene celebrates William of Normandy's invasion of Britain, whose soon-to-be-vanquished King Harold is seen on his throne deeply concerned about William's invasion craft.

of Halley's Comet before Halley's time. It appeared in the sky as William of Normandy was planning his invasion of England. The shrewd William promptly announced that the comet prophesied disaster for the English, and this helped to buoy Norman spirits. To be sure, Harold of England won a great victory over the Norse invaders in September of that year, but he then had to race southward to face the Normans and was disastrously defeated and killed in battle in October.

After William's success, a long strip of linen called the

Bayeux Tapestry was embroidered with 70 scenes of the victorious invasion. In one of them Halley's Comet is clearly depicted as a star with a tail, and the tapestry shows men pointing to it. Latin words state: "These wonder at the star."

Before 1066, European mentions of comets are very

The appearance of Halley's Comet in 1066 was very bright and fast moving through the heavens. (A nineteenth-century depiction.)

spotty and uncertain. In those times, however, astronomical observations in China were far in advance of those in Europe. In the 1700s these observations were brought back to Europe by returning Jesuit missionaries, and were first translated into European languages in 1846. These add to our knowledge of the early appearances of Halley's Comet.

September 9, 989. Mentioned in both European and Chinese records.

July 9, 912. Mentioned in both European and Chinese records.

February 27, 837. Mentioned in Chinese records only. It was a bad time for Europe, a time of civil war and Viking raids, and the quality of life was at a low ebb. The Chinese, however, listed no fewer than four comets that year, and, from the reported positions, the first of these was Halley's Comet.

May 22, 760. Mentioned in both European and Chinese records.

September 28, 684. There is not only a report of a bright comet for this year in the chronicles of the town of Nuremberg, Germany, but also included is a dramatic drawing of the comet and its tail. This is the oldest known pictorial representation of a comet.

March 13, 607. Possibly mentioned in Chinese records.

September 25, 530. Possibly mentioned in European records.

June 24, 451. Europeans noticed the comet this time, and with good reason. It was the year that the invasion of the western provinces of the Roman Empire by Attila the Hun was at its high point, and all of Europe trem-

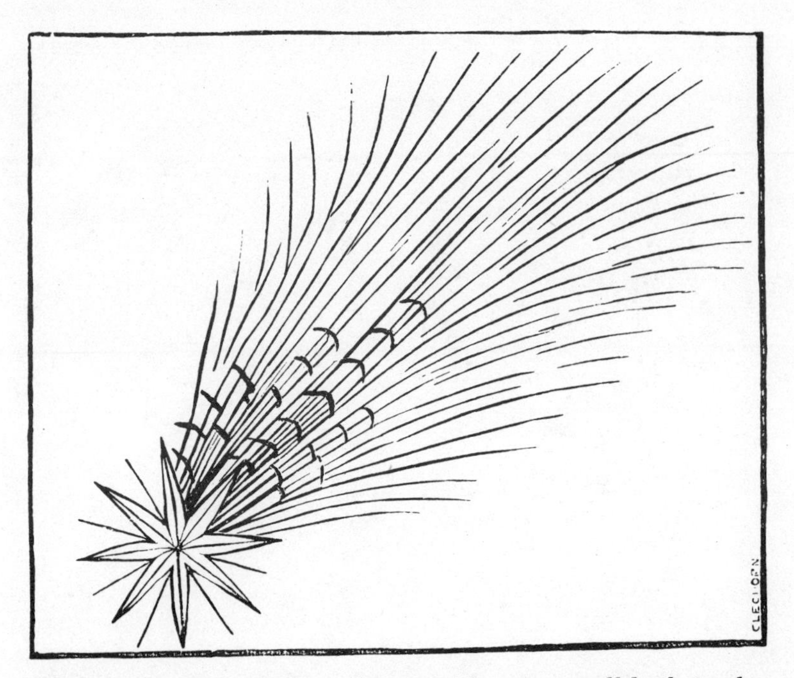

Halley's Comet of 684, as depicted in a woodblock in the Nuremberg Chronicle, was supposed to have presaged catastrophic storms, a pitiful harvest, and plague. This is the oldest known pictorial representation of Halley's Comet.

bled. However, at the Battle of Châlons in that year, Attila suffered the only defeat in his career. Two years later he died, and his Hunnish Empire fell apart. Halley's Comet may have predicted disaster for Attila, but it brought good news to Europe.

February 16, 374. Mentioned in the Chinese records.

April 20, 295. Mentioned in the Chinese records.

May 17, 218. Mentioned in the Chinese records.

March 20, 141. Mentioned in the Chinese records. For

This Dutch coin commemorated the return of Halley's Comet in 1577.

three centuries Halley's Comet is not mentioned in those European records of astronomy that have survived. It was the time of the Roman Empire. The Romans were not interested in science, and in those centuries the clever Greeks had grown to be interested only in philosophy and theology.

January 26, 66. Mentioned in the Chinese records. A passage in the writings of Josephus, the Jewish historian, may refer to it as an omen of the fall of Jerusalem to the Romans, which took place four years later.

October 5, 11 B.C. Mentioned in the Chinese records. It was also mentioned by a Greek writer who connected it with the death of Agrippa. However, Jesus was born somewhere about this time (the usual date given is 4 B.C., but it might have been somewhat earlier—the Bible

In 1910 this commemorative German coin depicted both Halley and his comet of that year.

A nineteenth-century English painting shows Julius Caesar and his wife, Calpurnia, presumably discussing a spectacular comet. The alarm on Calpurnia's face suggests that the comet portends evil things—particularly Caesar's murder on March 15, 44 B.C. (A comet was actually recorded in June of that year, three months after Caesar's death, and was regarded a sign of his ascension to heaven.)

gives no specific year). It may have been this appearance of the comet that inspired the writer of the Book of Matthew to tell the story of the Star of Bethlehem. Some astronomers who have worked out the appearance of the sky at the time of the Nativity feel that the tale of the Star of Bethlehem may have been inspired by a conjunction (a close approach in the sky) of the planets Jupiter and Saturn, and that it doesn't involve comets at all.

August 2, 86 B.C. No reports.

October 5, 163 B.C. No reports.

March 30, 239 B.C. Mentioned in the Chinese records.

316 B.C. No reports.

392 B.C. No reports.

467 B.C. Possible mention in European records. Before that, nothing

To be sure, not all famous comets represent returns of Halley's Comet. To cite just two examples: neither the Comet of 44 B.C. that was supposed to herald Caesar's assassination, nor the Comet of 1577, whose distance Tycho Brahe tried to determine, were Halley's Comet.

4

Dim Comets

The excitement about the return of Halley's Comet in 1759, and the realization that comets must be ordinary members of the solar system, did not reduce comet fear. Millions of uneducated and superstitious people still believed that comets were omens of disaster (as they believed many other varieties of nonsense), and do so to this day.

In addition, a new fear was introduced. Whereas planets moved in only slightly elliptical, and all but unchanging, orbits and always remained far apart, this was not true of comets.

Even though comets might be members of the solar system, with orbits of their own, these orbits were highly elliptical and carried them past all the planets, including Earth. To be sure, they usually passed these planets with plenty of room to spare, but planetary pulls were always changing their orbits slightly. What if one of these changes brought about a collision between a comet and the Earth?

In 1711, for instance, an English mathematician, William Whiston (1667–1752), published a book in which he endeavored to show that Halley's Comet had a longer orbit than Halley had thought, and that it returned to the vicinity of Earth not every 75 years, but every 575 years. Seven returns earlier, in 2345 B.C., it had made a very close approach to the Earth (he said), and at this approach its gravitational pull caused enormous tides, while the tail of the comet, striking Earth's atmosphere, caused a prolonged rain. The result was Noah's Flood. At a future return, Whiston predicted, the comet would cause the Earth to approach the Sun and be destroyed by fire.

In 1745 the French naturalist Georges L. L. de Buffon (1707–88) came up with another, but less frightful, collision theory. He suggested that about 75,000 years ago a comet had collided with the Sun. The matter that splashed out of the Sun from that collision solidified into the planets, including Earth.

As a result of speculations such as these, people who might never have dreamed of worrying about ill omens and astrological disasters began to worry about physical catastrophe and the possibility of cometary collisions.

As it happens, both Whiston's and Buffon's speculations have nothing to them, for reasons that will be given later. However, as I will explain in the last chapter, it *is* possible that comets might bring catastrophe to Earth, and they may even have already done so.

After the return of Halley's Comet, however, most astronomers had no time to worry about possible catastrophes, past or future. Instead, they grew furiously interested in comets, for these had suddenly become the hottest topic in astronomy.

Everyone wanted to discover a comet and if possible

work out its orbit and predict its return. After all, now that good telescopes existed, it was possible to detect comets too faint ever to be seen by the unaided eye. With a telescope, one can be pretty sure of discovering new comets quite frequently. Most of them are dim and unimpressive, to be sure—but a comet is a comet.

Messier, who had been the first professional to spot the returning Halley's Comet, grew to be totally absorbed in comet hunting. It was his only interest in life. In a half century of searching he discovered 21 new comets, and was chagrined whenever any other astronomer discovered one ahead of him. When his wife died he sincerely mourned her, but he did mention his unhappiness over the fact that sitting at her bedside had cost him a period of time during which he had been unable to hunt for comets.

To be sure, none of Messier's comets proved to be of any importance in themselves, but they made some things clear. It became obvious, for instance, that comets were very numerous, and this was another reason to see that it was silly to consider them omens of ill fortune, or of anything else. There were too many of them.

The matter of working out orbits and predicting returns proved to be not as easy as expected. Working out the orbit of a comet from a single passage across the sky was very difficult. Halley had had the good luck to be working with a bright comet, so he could trace its previous appearances, which had been well reported, and which gave him additional data that helped enormously in working out the orbit.

The numerous dim comets that later astronomers worked with, however, had not been observed previously as far as anyone knew, and attempts at working out orbits from one passage failed.

A French artist gave the comet arms to tear apart the world in this 1857 nightmare of a comet's collision with Earth.

Half a century passed after Halley's Comet's triumphant return in 1759, and still no other comet had had its orbit determined. In desperation some astronomers suggested that Halley's Comet might be a unique exception—that other comets did not have elliptical orbits, and therefore passed through the solar system only once.

There was a partial success in 1770, however, when a dim comet was observed by a Swedish astronomer, Anders Johan Lexell (1740–84). He managed to calculate its orbit and found it to be so small that the comet returned every 5.6 years. Unfortunately, Lexell's Comet, as it was eventually called, after it had made its trip about the Sun, was never seen again.

In later years its orbit was followed backward and forward from 1770, and astronomers discovered what had happened. Originally Lexell's Comet had had a long orbit, but in 1767 it had passed near Jupiter, whose gravitational attraction pulled it into its 5.6-year orbit. It passed the Sun and returned in 1776, but was so dim that apparently no one noticed it. Then, on its way outward in 1779, it passed near Jupiter again and its orbit

This artist, poking fun at the comet-gazing craze of the nineteenth and early twentieth centuries, shows the "discovery" of a comet at Greenwich Observatory in 1906.

Morehouse's Comet appears as a dim comet with a long, tenuous tail in this 1908 telescopic photograph.

was changed into such a long ellipse that it coud be expected to return to Earth's vicinity rarely, if at all. Even if seen again, it would be considered a new comet, for there would be no way of identifying it as Lexell's Comet.

But in 1802 things changed. The German mathematician Karl Friedrich Gauss (1777–1855) devised a mathematical method for calculating an orbit from as few as three well-separated observations. With that, it became much easier to do the job.

When in 1818 a French astronomer, Jean Louis Pons (1761–1831), discovered a new comet, the German astronomer Johann Franz Encke (1791–1865), who had been a pupil of Gauss, set about working out its orbit. He succeeded in 1819, and the comet has been known as Encke's Comet, or Comet Encke, ever since. He found that it had a small orbit and that it circled the Sun in only 3⅓ years. This is, in fact, the smallest cometary orbit that is known to this day, and Encke's Comet has been spotted on every one of its almost 40 returns since 1819.

Encke's comet was only the second comet to have its orbit calculated and its returns observed, Halley's Comet having been the first. It took 114 years after the first cometary orbit was determined before the second came along, but after that there was no difficulty. Numerous

This artist shows Chinese villagers trying to frighten away Halley's Comet in 1910 with torches and bonfires.

Halley's Comet was photographed telescopically as it approached perihelion in 1910.

comets had their orbits established; Encke himself calculated no fewer than 56 such orbits.

All comets, astronomers now believe, are part of the solar system and have orbits that take them about the Sun in longer or shorter periods.

Comets can be divided into "short-period comets" and "long-period comets." Short-period comets have periods of revolution of less than 200 years; Halley's Comet is a short-period comet.

Long-period comets have orbits so elongated that it is exceedingly difficult to work out their orbits from the short section near the Sun during which the comet is visible. It may take such comets thousands, hundreds of thousands, even millions of years to make one circuit about its enormous orbit.

Such comets, even though often very bright and spectacular, may never be seen twice by astronomers. After all, the last time they passed near Earth there was no one to observe them intelligently, since, at best, only the primitive ancestors of modern human beings roamed the Earth. The next time they return, who knows what the fate of humanity might have been, and whether any of our descendants will be left on Earth to observe.

5

The Death of Comets

In 1806 the German astronomer Friedrich Wilhelm Bessel (1784–1846) discovered a comet. He searched the records, studied passages across the sky, and decided that it might be the return of a 1772 comet that was one of those discovered by Messier. Later on, Bessel decided he had made a mistake in his calculations and was wrong, but others thought he had *not* made a mistake and was right. The dispute fixed astronomical attention on the comet.

An Austrian army officer, Wilhelm von Biela (1782–1856), who was also an amateur astronomer, decided to watch for the comet, which was supposed to appear in 1826 if Bessel's suggestion was correct. On February 27 he spotted a comet and followed it through the sky for twelve weeks. He then calculated its orbit—which was now easy, thanks to Gauss—and found that its period of revolution was 6 years and 9 months.

Calculating backward, he proved that Bessel had indeed been right all along. The comet, which came to be known as Biela's Comet, had appeared in 1772, had passed through its next four returns without being noticed, since it was a dim comet, and then in 1807 had been spotted by Bessel. Two more unnoticed returns took place, and then it was seen by Biela and its orbit was calculated.

Because of the controversy, Biela's Comet became widely known and a number of astronomers took to calculating its orbit very carefully, taking into account various planetary pulls to make sure that they would have an exact idea of when it might return. They did not want to waste their time if it was going to suffer the fate of Lexell's Comet and be hurled, by planetary influences, into a new orbit, or out of the solar system altogether.

One German astronomer, Heinrich Wilhelm Olbers (1758–1840), calculated that, on its next return in 1832, Biela's Comet would pass very close to Earth's orbit on October 29 as it was flying toward its perihelion. To be sure, Earth itself would not be in that part of its orbit but in another part, fifty million miles away—but somehow people didn't pay attention to that at first.

A widespread feeling grew that there was going to be a collision on October 29, 1832, and a comet panic started to build. But astronomers hastened to explain over and over again that there would be no collision and, amazingly, the panic died down. Biela's Comet did pass quite close to Earth's orbit when the time came, just as Olbers had predicted, but, of course, nothing happened to the distant Earth.

Halley's Comet returned again and passed perihelion on November 16, 1835, but having just gone through a near-panic over nothing, people took the appearance with a certain calm, especially as it was one of the times when the comet was not particularly spectacular.

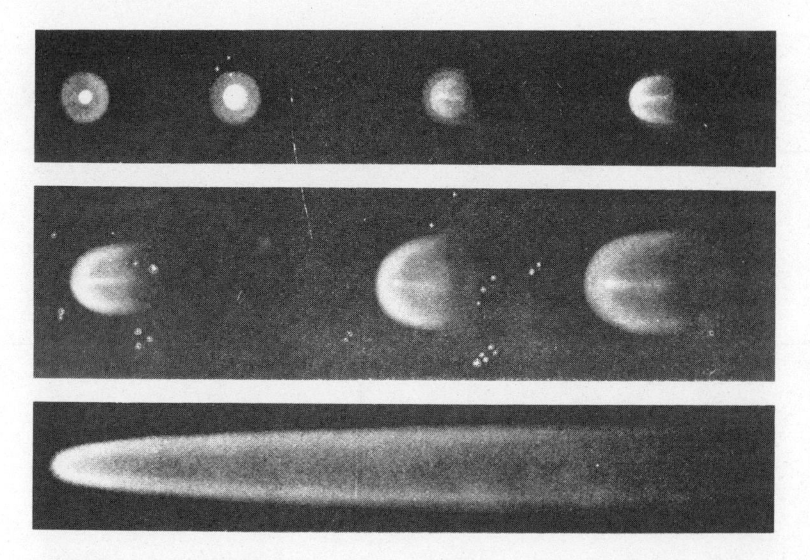

The head of Halley's Comet and the growth of its tail, as drawn by an astronomer, is shown at various points as it approached perihelion in 1835–36. Features of comets often appear more detailed when drawn by an observer using a telescope than in photographs taken through a telescope.

On November 10, when Halley's Comet was nearly at perihelion, a child was born in the little town of Florida, Missouri. He was Samuel Langhorne Clemens, who was to live to become (in my opinion) the greatest of American writers, better known under his pen name of Mark Twain.

After Halley's Comet disappeared for another three-quarters of a century, Biela's Comet continued to be of particular interest to astronomers. It appeared again in July 1839, and then once more in February 1846. The first to see it on this later occasion was an American oceanographer and astronomer, Matthew Fontaine

Maury (1806–73). He reported that two comets were moving along side by side, each with its own tail. Clearly, Biela's Comet had split in two.

In 1852, when it returned once again, the first to see it was an Italian astronomer, Pietro Angelo Secchi (1818–78). The two parts of Biela's Comet were now widely separated, and one was a little ahead of the other. There was no chance to see it again in 1859 because it was in the sky during twilight hours and was too dim to see when the sky was not entirely dark.

It should have appeared in 1866 under conditions that would have made it clearly visible, but it didn't. In fact, Biela's Comet was never seen again, even though it had not approached a planet in such a way as to have its orbit changed. It had just crumbled away and, so to speak, died. (Since then, other comets have also split up and died.)

The fate of Biela's Comet made it seem that comets were perhaps rather light and even insubstantial objects, nothing at all like planets. This helped allay fears of the possibility of collisions causing planetary floods and other catastrophes. In particular, Buffon's notion of the effects of a cometary collision with the Sun is very wide of the mark. Actually, a comet occasionally *does* collide with the Sun. The comet is then destroyed, but the much vaster Sun is not affected in any noticeable way.

A number of suggestions were made as to what the structure of comets might be, in order to account for their appearance and their fragility. The one that is accepted today by nearly everyone is that advanced in 1950 by the American astronomer Fred Lawrence

A British astronomer carefully drew the head of the Great Comet of 1861 as he saw the nucleus and surrounding coma in July of that year.

Whipple (b. 1906). It is referred to informally as the "dirty snowball" theory, and I remember his explaining it to me at a dinner party shortly after he had advanced it.

Whipple suggested that a comet is essentially a ball of icy substances (a "snowball") that, at higher temperatures, would turn into vapors. The frozen substances that might make it up include water, of course, and also ammonia, methane, carbon dioxide, hydrogen cyanide and so on. Embedded in the icy substances would be dust particles and little bits of solid, rocky materials. (These would make the snowball "dirty.") At the core there might be a solid bit of rock, or the comet might be dust and grit embedded in ice all the way through. The frozen ball might be several miles across and would be a small and barely noticeable object as long as it remained frozen.

While the comet is at the far end of its orbit, it does stay icy and hard, and it is too far from Earth to be seen in any case. As it approaches the Sun, however, the comet's temperature rises. Some of the ice evaporates and the dust it contains is liberated. The core of the comet, whether solid rock or still-frozen ice, might shine as a star-like point called the "nucleus," and around it there would be the haze of dust, which is the "coma." The coma would grow larger and larger as the comet approached the Sun, until in a particularly large comet it might expand to the volume of a major planet. All that volume would be filled with insubstantial dust, however. The coma would sweep into a tail, which might stretch over hundreds of millions of miles, though all the matter in it, if it could be gathered together, would fit into an ordinary living room, or even a suitcase.

Halley's Comet moved through the sky with a diminished tail as it departed from perihelion on June 6, 1910.

Once the comet passes the perihelion point and begins to recede, it starts to cool. The tail shrinks, the coma collapses, and finally the comet is just a hard ball of ice again.

Each time the comet circles the Sun, however, some of its material is gone forever. The material in the tail spreads through space and never returns to the comet. The same is true of much of the matter in the coma. Every time the comet returns, therefore, it is smaller than before and becomes less and less spectacular.

That is why short-period comets are so dim. They have made many returns and have dwindled away. Sometimes they dwindle away to nothing, as Biela's Comet did. Sometimes a rocky core remains behind, as seems to be true of Encke's Comet, which still produces a small amount of haze each time it nears the Sun, but which doesn't seem to change much from passage to passage.

Because Halley's Comet approaches the Sun once in 77 years, it has made only 32 returns since the time of ancient Greece's golden age, 2500 years ago. In that same time, Encke's Comet has returned over 750 times. Besides that, Encke's Comet approaches the Sun more closely at perihelion than Halley's Comet does, so that the former is more strongly heated and more extensively evaporated.

That is why Encke's Comet is practically dead, while Halley's Comet is still very much alive. Even so, Halley's Comet can't be quite the spectacular object it once was. And the time will come, several thousand years hence, when it will surely be reduced to a dim object visible only in telescopes, if it hasn't crumbled away and died altogether.

A close-up of the head of Halley's Comet on May 9, 1910.

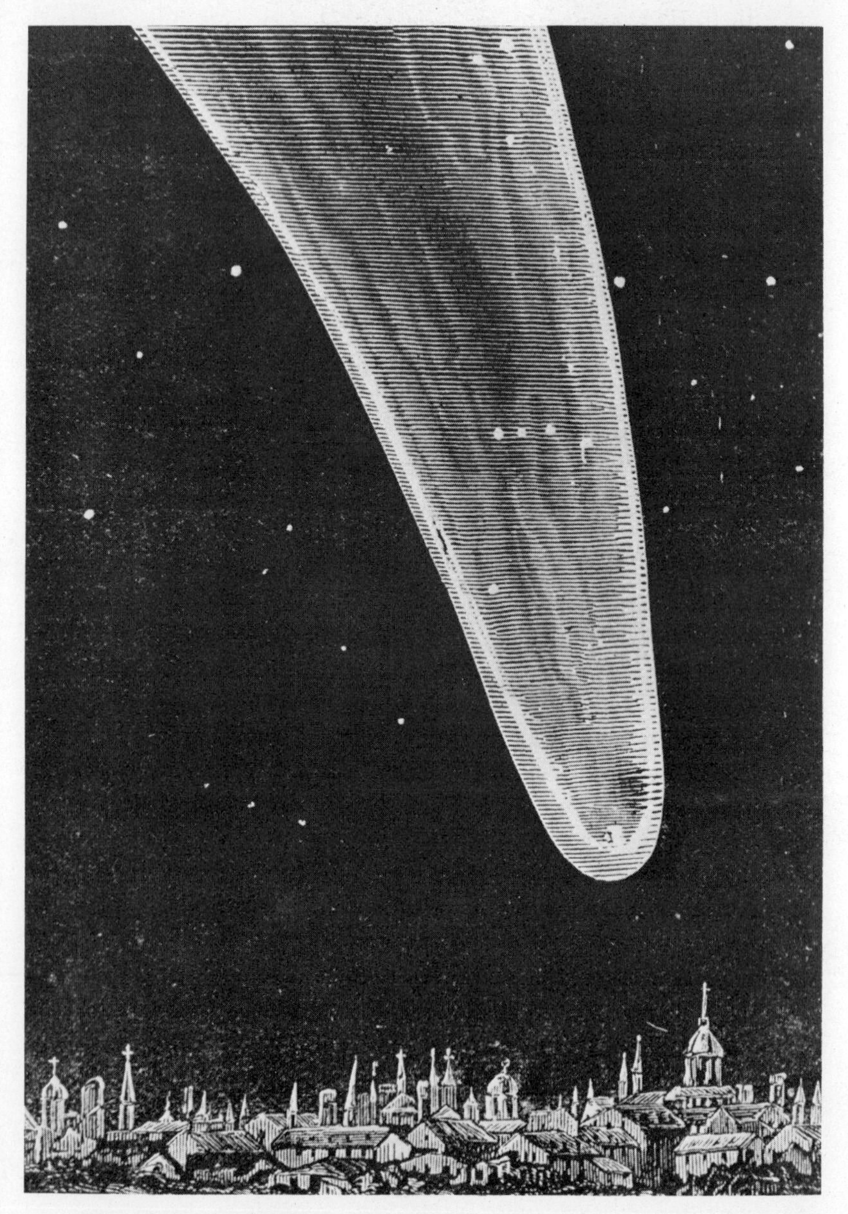

The first of several spectacular comets in the nineteenth century was this giant finger of light, the Great Comet of 1811.

6

Nineteenth-Century Comets

Naturally, the most spectacular comets are the ones with very long, long orbits that approach the Sun only once in many thousands of years. They have had only a few occasions to lose any of their substance, and so they grow huge comas and spectacular tails as they approach the Sun. Sometimes they are called "new comets" because they have had so little chance to undergo change, and because, when they do approach, it will be the first time they will ever have been observed scientifically.

In the nineteenth century, a number of such new comets blazed across the sky.

In 1811, for instance, a huge comet appeared in the sky, with a tail that grew to be about 100 million miles long (longer than the distance from the Sun to the Earth). It remained visible for a year and a half and was very bright for many weeks. As it happened, Portugal produced a very good vintage of port wine while this comet was in the sky. The comet had no connection whatever with the wine, but the vintage was advertised and sold as "comet wine" for over half a century.

In former centuries a comet that spectacular would have sent Europe mad with fear over approaching disaster, and there must have been many who were ill at ease as it was. The Emperor Napoleon was apparently not concerned over the comet, and it was the worse for him that he wasn't. The comet had barely disappeared when he launched his Russian invasion, one which ended in total disaster for him and made his downfall inevitable.

In 1843 a comet appeared that was probably even brighter than the one of 1811, but it was poorly placed for visibility in Europe. Its tail was straight and narrow and extended over one-quarter of the sky.

The Comet of 1843 was remarkable in that it approached the Sun very closely as its perihelion. At its closest approach, Halley's Comet is 55 million miles from the Sun, while Encke's Comet is only 31 million miles from the Sun. The Comet of 1843, however, was only 500,000 miles from the Sun's center when it reached perihelion. It skimmed the surface of the Sun at a distance of only 80,000 miles or so. The closer an object is when it circles the Sun, the faster it moves, and the Comet of 1843 was moving so quickly at perihelion that it passed three-quarters of the way around the giant Sun in less than a day, moving at a rate of up to 350 miles a second. It was that speed that helped it survive the encounter.

The Comet of 1843 belongs to a class of comets known as "Sun-grazers," and Halley's Comet is not one of them. If it were, two or three passes would have been enough to vaporize it to a rocky core, or to nothing at all.

On June 2, 1858 the Italian astronomer Giovanni Battista Donati (1826–73) detected a comet that has been

The Great March Comet of 1843, shown here over Paris, was a Sun-grazer; it whisked around the Sun at 1,270,000 miles per hour and passed only 80,000 miles from the solar surface.

known ever since as Donati's Comet. It was the third new comet of the century. More important than its brightness, however, was the fact that it had several tails and that these changed shape from time to time. There were also distortions in the coma.

Observations of Donati's Comet made it clear that as the cometary surface heated up, gouts of gas could be given off. From the modern theory of comet structure, this is not surprising. It is possible that crusts of rocky materials may enclose ice. As the ice vaporizes, the accumulating vapor would eventually blow the rocky material away in an explosion that would have an effect on

The Italian astronomer Giovanni Donati detected a comet in June of 1858 that would become very bright and feature a dust tail and one or more slender gas tails.

the shape of the coma and on the tail. These explosions act like rockets, pushing the comet forward, backward, or sideways. They are bound to change the orbit slightly even when there are no planets close enough to exert perceptible pulls. It is for this reason that even when all planetary influences are taken into account, comets are still likely to reach perihelion a little before or a little after the calculated time.

In 1864 Donati was the first to manage to get a spectrum of a comet; in other words, to analyze the wavelengths of the light it gives off. The spectrum possessed dark lines, places where no light of that particular wavelength was given off. Such lines mean that certain substances surrounding the comet have absorbed light, and from the position of the dark lines it is possible to tell the chemical nature of those absorbing substances. In 1868 the English astronomer William Huggins (1824–1910) was able to identify some of the substances in the coma. That was the first step toward the theory of cometary structure Whipple developed over eighty years later.

In 1861, meanwhile, a fourth new comet appeared. It was first detected deep in the southern hemisphere, but the Australians who saw it had no way of communicating with Europe or the United States except by ship. By the time the news reached its destination, the comet had moved into the part of its orbit that made it visible from the northern hemisphere, so that it burst upon Europeans and Americans without warning. Like Donati's Comet, the Comet of 1861 showed a tail of complicated and changing structure.

It was a particularly large comet in appearance because it came unusually close to the Earth. At its closest it was only 11 million miles away, or less than half the minimum distance of Venus, the planet that is nearest us. About June 30 the comet's tail swept over the Earth. Nothing happened as a result, of course, since the tail, for all its impressive appearance, is so insubstantial. There were a number of Americans, however, who later insisted that the Comet of 1861 was sent by God as a warning of the coming carnage of the American Civil

War (as though by that time one needed a comet to foresee the tragedy).

The fifth and last new comet of the nineteenth century appeared in 1882. It split off a small fragment as it traveled and that fragment grew fainter and disappeared—another sign of the fragility of comets. The Comet of 1882 followed the same orbit as the earlier Comet of 1843, and was also a Sun-grazer. The two could not have been the same comet, however: the Comet of 1843 couldn't possibly have returned after only 39 years.

It is clear now that there are a whole family of Sun-grazers following the same orbit. They may represent a single large comet that broke up into numerous fragments at the previous passage very close to the Sun, many thousands of years ago. The fragments are returning in a string, some having slowed a bit and some having speeded up as the result of explosive effects on the surface of the comet.

The Scottish astronomer David Gill (1843–1914), serving in an observatory in South Africa, photographed the Comet of 1882, and this was the first good photograph taken of a comet. In 1894 the American astronomer Edward Emerson Barnard (1857–1923), took a telescopic photograph of a patch of sky and discovered an unknown comet within the patch. Since then, more and more discoveries have been made through photography and fewer and fewer by eye (with or without a telescope). Thanks to photography and other advanced techniques, some ten to thirty comets are discovered every year now.

The five new comets of the 1800s, appearing within the space of 71 years, all had periods in the thousands

The Great Comet of 1860-61, shown here, had a magnificent tail that swept over the Earth (without noticeable effect) in late June of that year. In the United States, this comet was thought to foretell the carnage of the Civil War.

The Great September Comet of 1882, another Sun-grazer, shone brightly during an eclipse of the Sun on May 17, 1882.

of years, perhaps many thousands of years. The fractions of their enormous orbits during which they remained visible were too small to calculate accurately at the time. Just as there was no way of predicting their coming, there is no way of predicting, even approximately, when they will return.

In the century since 1882, there have been no really spectacular new comets of this type—not one. In fact, the brightest comet visible in the skies of the northern hemisphere since 1882 appeared in 1910, and it was Halley's Comet once again, as it made its third return since Halley had worked out its orbit.

Of course, what with telescopes and photography, Halley's Comet could be detected long before it became visible to the unaided eye, and it remained detectable long after it ceased being visible to the unaided eye. It was first photographed on September 11, 1909, and was followed photographically until July 1, 1911, at which time it was beyond Jupiter's orbit.

Halley's Comet passed its perihelion on April 20, 1910, and was three days late in doing so, despite the fact that every possible gravitational effect had been allowed for. Undoubtedly the lateness was due to the rocket effect of the explosive heating of its surface.

Even in 1910 there was plenty of comet fear. In particular, it seemed quite likely that Earth would pass through the tail of Halley's Comet. Although astronomers assured the world that this would amount to nothing (as had been true in the case of the Comet of 1861), many people were convinced that it would bring about the end of life on Earth, if not the end of the planet itself. Some unscrupulous salesmen made considerable money by selling "comet pills" that were advertised as antidotes for the effect of the poisonous gases supposed to be released into the atmosphere by the tail of Halley's Comet.

Needless to say, the Earth and its burden of life experienced no effects whatever from the tail of Halley's Comet.

In 1910 Mark Twain lay dying. When those about him spoke with hope, he shook his head. "I came with the comet," he said, "and I shall go with it." He died on April 21, the day after it passed perihelion.

7

Comet Tails and Meteors

The first scientific discovery made about comets was that their tails always pointed away from the Sun. But why?

When something on Earth is moving rapidly along on a windless day and is emitting smoke as it does so (a steam locomotive, for instance), we expect the smoke to trail behind as it moves. That is because air resistance has a far greater effect on the tiny smoke particles than on the massive train. The smoke is therefore slowed and lags behind the train.

In a vacuum, however, smoke emitted by a moving object would move right along with the object. It would not lag behind.

When a comet is approaching the Sun, the dusty material in the coma streams backward and lags behind the comet; yet the comet is moving through a vacuum. What's more, as the comet swings around the Sun, the tail swings about too, so that at every moment it extends outward in the direction away from the Sun.

When the comet is halfway around the Sun, the tail

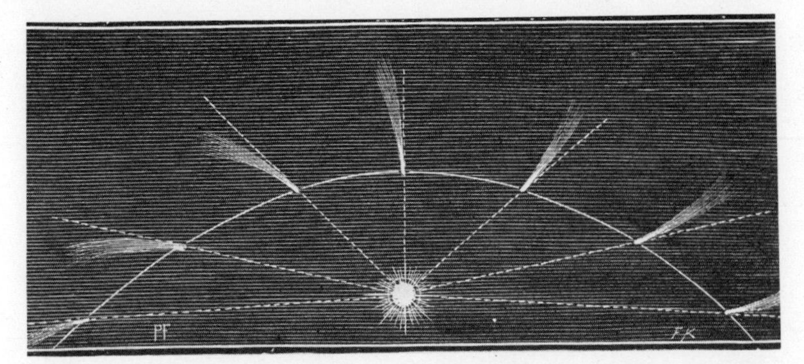

A comet's tail always points away from the Sun at every point in its orbit. When a comet has rounded the Sun and is returning to the other end of its orbit, its tail flows in the direction of its motion—"blown" by the solar wind.

still points away from the Sun. It is then at right angles to the comet's motion.

Once the comet has completed its turn about the Sun and is moving away toward the farther portions of its orbit, its tail *still* points away from the Sun, and actually moves *ahead* of the comet.

Since we are so used to motion through atmosphere, this seems a strange phenomenon to us. We would find it peculiar indeed to have a steam locomotive race along the tracks with the smoke from its funnel moving at right angles to the locomotive's motion, or perhaps streaming along ahead of the train. The only way we can imagine that happening would be to suppose that there was a strong wind moving at right angles to the locomotive, or moving in the direction of the locomotive at a greater speed.

Could it be, then, that there is some kind of "wind" issuing from the Sun that succeeds in "blowing" the tail?

We do know that light and other radiation blazes out from the Sun in every direction. The Scottish physicist James Clerk Maxwell (1831–79), in his theoretical studies, pointed out that such radiation ought to exert pressure so that a stream of it could act as a very feeble wind.

In 1901, using very light mirrors suspended in a vacuum, the Russian scientist Peter Nicolaevich Lebedev (1866–1911) actually measured the pressure of radiation, and proved that it really existed. For about half a century thereafter, such radiation pressure was considered to be responsible for the behavior of comets' tails. Radiation pressure was continually pushing the dust of the coma into a tail pointing away from the Sun, dissipating it gradually through the vast empty reaches of space.

It eventually turned out, however, that radiation pressure, although it existed, simply wasn't strong enough to account for comets' tails. Was there anything else the Sun gave off?

During the 1920s an English scientist, Edward Arthur Milne (1896–1950), made careful theoretical studies of the behavior of the Sun's atmosphere. He calculated the gravitational force pulling it inward and the radiation pressure pushing it outward. It seemed to him that at the surface of the Sun the radiation pressure was so strong that it could push particles away from the Sun at enormous speeds—even against the pull of gravity.

The most common particle making up the Sun is the hydrogen atomic nucleus, which is an electrically charged proton. Milne predicted, then, that there must be a stream of electrically charged particles emerging from the Sun in all directions.

By the 1950s scientists were studying the regions of space beyond Earth's atmosphere with rockets, and the Italian-American physicist, Bruno Rossi (b. 1905) showed that there were indeed fast-moving electrically charged particles streaming away from the Sun and passing the Earth. This is now called the "solar wind."

Astronomers are now convinced that it is this solar wind that sweeps the coma of comets into long, dissipating tails, and it is no longer a surprise that the tails always face away from the Sun and precede the comets as they recede from the Sun. The solar wind moves much faster than the comets, overtakes them, and sweeps the dust surrounding the comet ahead of it.

What happens to all the material driven away in the tail? Does it just vanish?

Nothing ever vanishes, of course. The vapors in the tail spread through space as individual atoms or molecules. They might be detected there in very subtle ways, but they wouldn't affect us. However, what about the little rock fragments that make up the coma along with the vapors? What happens to them?

They spread out behind and in front of the comet; more and more of them as more and more of the comet is vaporized in successive approaches to the Sun.

After the comet vanishes altogether, as Biela's Comet did, the small rocky fragments would eventually spread over the entire orbit. This would come about through the action of the solar wind and through the pull of the planets upon the fragments as they pass. Still, a particularly large concentration of them would remain near the spot where the comet itself would have been.

We have evidence of such rocky bits.

Anyone watching the sky on a dark, moonless night is

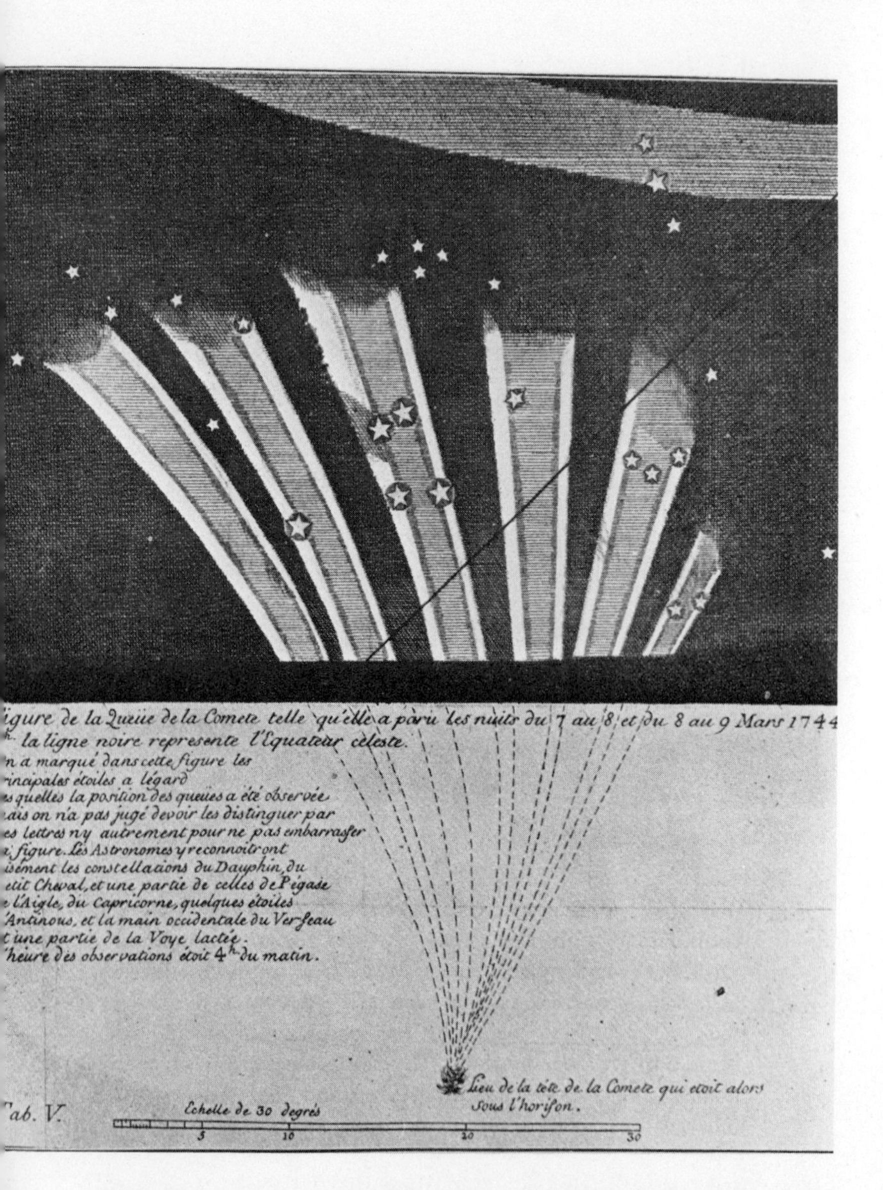

The Swiss astronomer Cheseaux diagramed the tails of the Great Comet of 1744. In March, the head of the comet was well below the horizon for European observers.

On a March night in 1744, the six tails of Cheseaux's Comet were observed and recorded by an artist.

bound to see tiny streaks of light now and then that last just a second or so. The first thought of children, or of unsophisticated adults, might be that a star had slipped from its place and fallen, and, in fact, these streaks of light are sometimes called "shooting stars."

They can't be stars, however, since no matter how many of them appear, no stars are ever missing from the sky as a result. Aristotle guessed that they might be simply an atmospheric phenomenon, flashes of light in the upper atmosphere, and in time he turned out to be right. Such shooting stars are therefore called "meteors," from Greek words meaning "high in the air."

In early times there were occasional reports of rocks, or lumps of iron, falling from the sky. As modern science developed after 1600, scientists became rather skeptical of such stories. They sounded unbelievable.

The first scientist to consider the tales of objects falling from the sky in a serious way was a Swiss naturalist, Johann Jakob Scheuchzer (1672–1733). In 1697 he suggested that the falling stones might have something to do with meteors.

The suggestion was disregarded, but there were more and more tales of falling stones, and a century later the German physicist Ernst Florens Friedrich Chladni (1756–1827) began to look into the matter. He actually collected stones said to have fallen and studied them. In 1794 he published a book in which he argued that small bits of matter in space would occasionally collide with the Earth. As these bits of matter streaked through the air, air resistance would heat them till they glowed white-hot and began to vaporize. The streak of white-hot matter was the meteor. The unvaporized portion that struck the ground were the objects that supposedly fell from the sky.

Chladni's explanation made so much sense that the scientific world began to waver. In 1803 the French physicist Jean Baptiste Biot (1774–1862) investigated the claim of a fall of thousands of fragments in northern France. His careful report settled the matter. Rocks and lumps of iron *did* fall from the skies. The fallen meteors were called "meteorites."

Not all meteors, however, give rise to meteorites. For instance, in November 1833 there occurred what is called a "meteor shower." Astonished onlookers in New England watched the night sky turn into endless streaks of light that were as thick as snowflakes. They came literally by the hundreds of thousands. Some of the people who were watching thought every star in the sky had fallen (as predicted in the biblical Book of Revelation) and that the end of the world had come. Nevertheless, the next day dawned exactly as usual and the next night saw the sky as full of stars as ever.

Most meteors are streaks of light produced by tiny bits of grit, no bigger than grains of sand. They vaporize totally while still high in the air and there is nothing left to reach the ground. On the morning after the great meteor shower of 1833, not one meteorite was found anywhere.

The streaks of light on that night in 1833 had all seemed to radiate outward from a point in the constellation Leo. These particular meteors were therefore referred to as the "Leonids."

In 1834 the American scholar Denison Olmsted (1791–1859), who had witnessed the shower, offered an explanation that has since been accepted.

The Leonids are, he said, a swarm of sand grains, so to speak, orbiting the Sun. Every year the Earth passes through the swarm and meteors appear in numbers greater than usual. In the case of the Leonids, the Earth passes through a particularly thick region of the swarm every 33 years, and there is then a meteor shower (though no meteor shower, before or after, has been as spectacular as that of 1833).

There are other meteor showers with the streaks radiating out from a point in one constellation or another. These are all named after the constellations from which they seem to emerge, so that you have the Perseids, the Lyrids, the Aquarids, and so on.

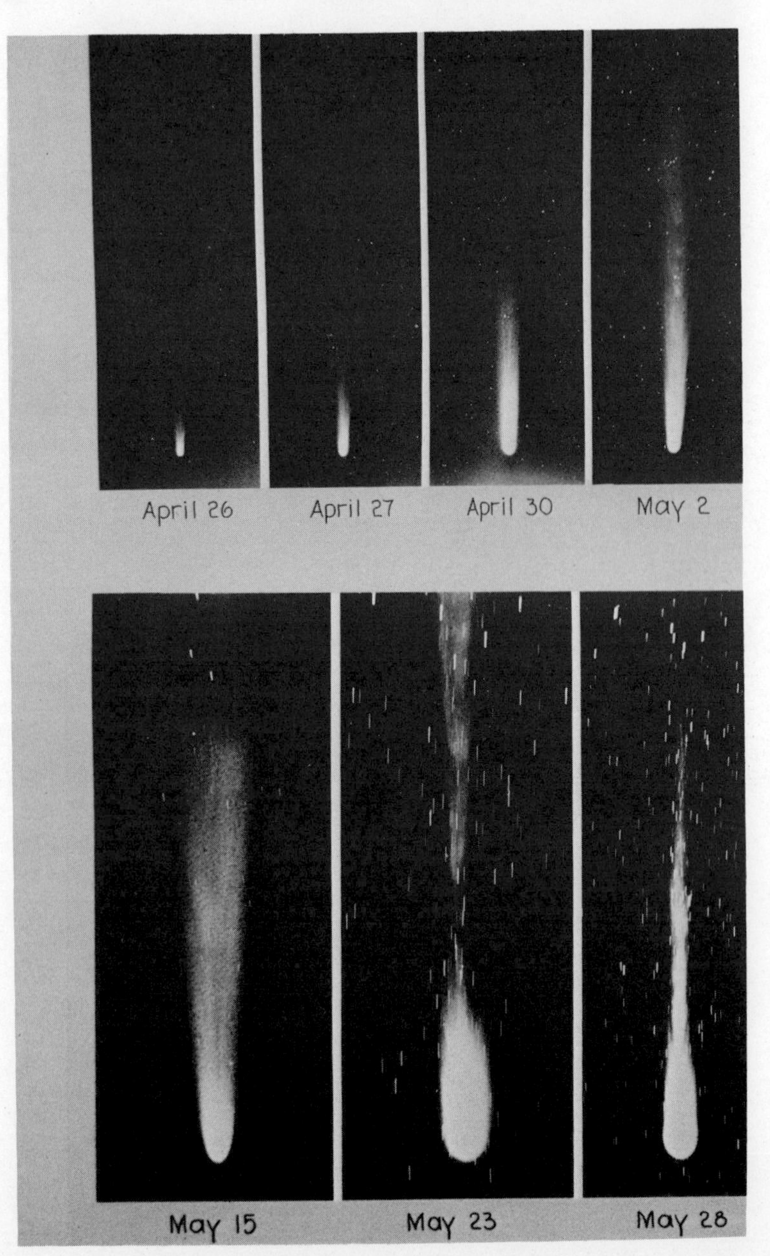
April 26
April 27
April 30
May 2
May 15
May 23
May 28

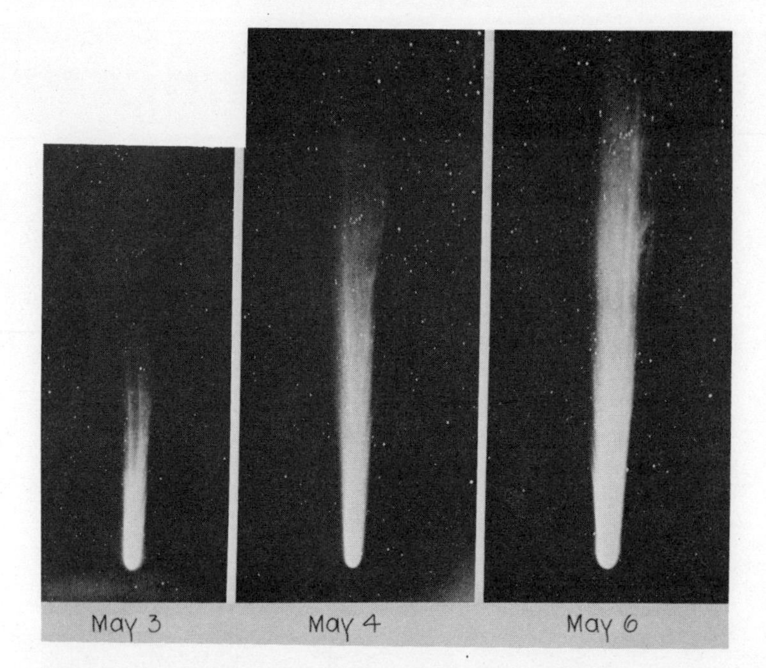

These telescopic photographs show the apparent waxing and waning of the tail on Halley's Comet in 1910.

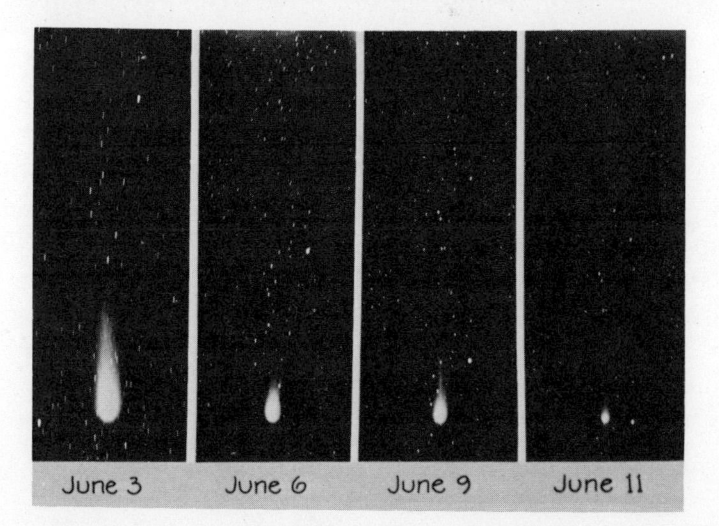

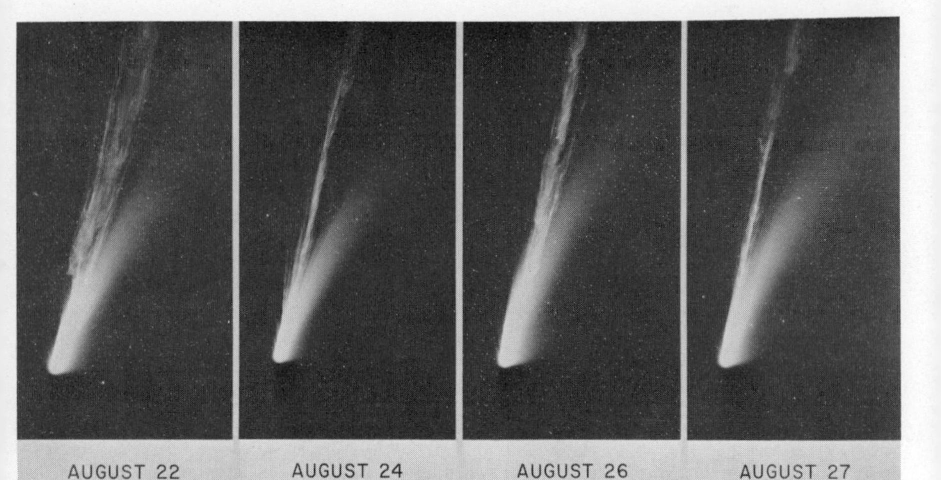

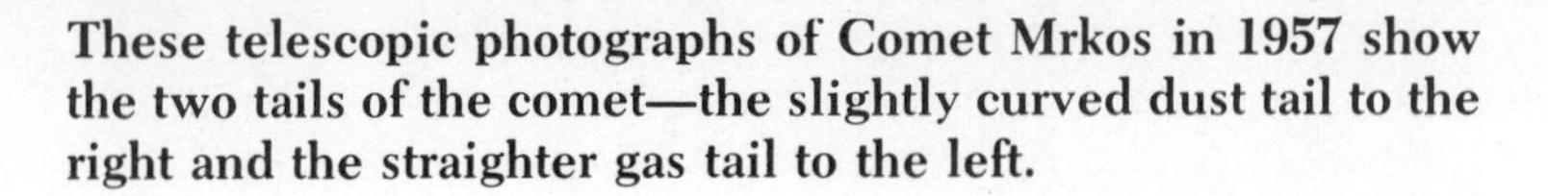

These telescopic photographs of Comet Mrkos in 1957 show the two tails of the comet—the slightly curved dust tail to the right and the straighter gas tail to the left.

In 1861 the American astronomer Daniel Kirkwood (1814–95), thinking perhaps of Biela's Comet, which by that time was clearly seen to be breaking up, suggested that all the particle swarms through which the Earth passed at various times were the debris of dead comets and were moving in what had once been their orbits.

The next step was to determine what the orbits of these meteor swarms actually were. In 1866 the Italian astronomer Giovanni Virginio Schiaparelli (1835–1910) was able to show that the Perseid meteor swarm did indeed have a comet-like orbit. What's more, it was an orbit that matched that of a comet that had been detected and followed in 1862.

Soon afterward, the French astronomer Urbain Jean Joseph Leverrier (1811–77), and the English astrono-

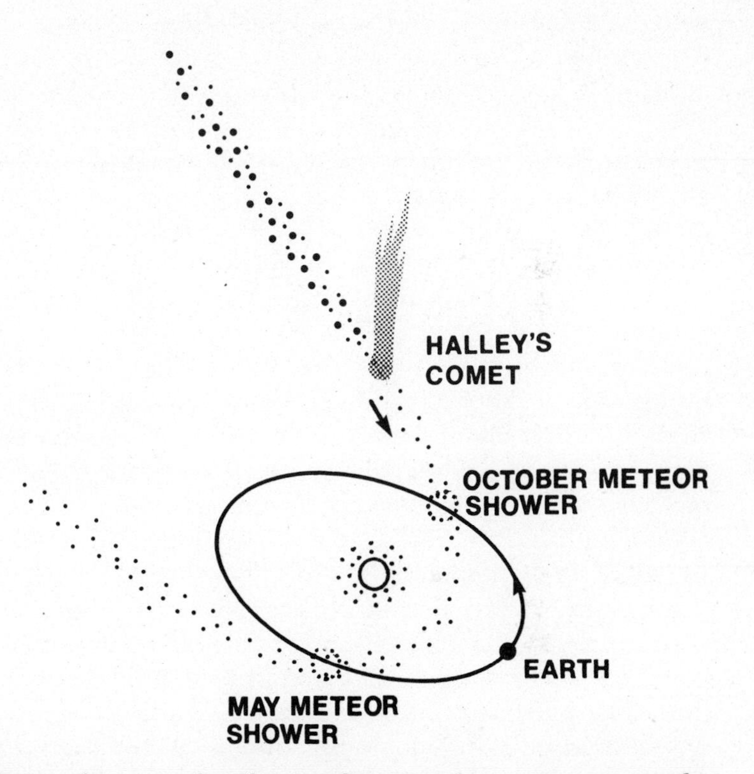

Meteor showers take place at the same time every year as the Earth passes through the orbiting debris trailing a comet's path. Shown here are the meteor showers resulting from the passage of Halley's Comet around the Sun: the Aquarid showers in October and the Orionids in May.

mer John Couch Adams (1819–92) independently worked out the orbit of the Leonid meteor swarm and found it to have a comet-like orbit as well. The Aquarid meteor swarm moves in the orbit of Halley's Comet, and is evidence of the slow disintegration of this most famous of all comets as a result of its passages around the Sun.

There was some question at first as to whether comets broke up into meteor swarms or were formed out of meteor swarms; in other words, which came first. Schiaparelli believed the meteor swarms came first.

The matter was settled by an Austrian astronomer, Edmund Weiss (1837–1917). He showed that the Andromedid follows the orbit of Biela's Comet, which had broken up in the 1860s. On November 27, 1872, when Biela's Comet should have appeared if it had existed, there was instead a heavier than usual shower of Andromedids. Clearly, that shower was the grit into which the comet had disintegrated, and Weiss gave the meteor swarm the new name of "Bielids."

Naturally, every time the Earth passes through a meteor swarm it collects millions of particles that are subtracted from the swarm. If a comet is still in existence, those lost particles can be replaced by further cometary disintegration. If, however, the comet has been reduced to a rocky core, or has disintegrated altogether, the particles swept up by the Earth and by other larger bodies of the solar system cannot be replaced, so the swarm gradually dwindles. Over the years, for instance, Bielids have grown fewer and fewer, until now they seem to be just about gone.

If the Earth is constantly passing through such meteor showers, which are the debris of comets, might it

In 1833 a meteor shower—as here at Niagara Falls—lit up the Northeastern states. It poured meteors into the night sky at a rate of 70 a second, or 250,000 an hour!

not just as easily hit a comet itself, since the orbits of swarms and of comets are so similar?

The two are not the same thing. The meteor swarms spread out through the cometary orbit and to all sides of it, so that bits of grit occupy wide stretches of space. The comet itself is confined to one spot in the orbit. Hitting a meteor swarm is like hitting the side of a barn; hitting a comet would be like hitting a small knothole in the side of the barn.

Of course, a comet has a tail that takes up an enormous volume of space, and the Earth can (and on occasion does) pass through the tail, but the matter in that tail is so thinly spread that it has no visible effect on us.

Nevertheless, even though a collision between a comet and the Earth is very unlikely, it is not entirely impossible. We will return to this subject later in the book.

8

The Distant Comets

How many comets are there? Kepler, when asked that question four centuries ago, answered, "As many as there are fish in the sea."

That was just a guess on his part. He had no evidence for the remark, for up to his time the total number of comets reported in the various surviving records throughout human history was less than 900.

Nowadays, of course, we see many more comets than people did in Kepler's time and before, for now we have the telescope, so that a new comet is spotted on the average of every two or three weeks.

Nevertheless, the comets we now know must be but a small fraction of all the comets there are. The largest comets have long orbits that extend far, far beyond the family of planets, and the vast spaces at that great distance may be the original home of the comets.

A good idea of how faraway comets might recede came in 1973, when a Czech astronomer, Lubos Kohoutek, detected an approaching comet while it was still beyond the orbit of Jupiter. Because it could be seen so far, it seemed to be a large comet, and therefore a new one coming from far out in the depths of space. It was followed all the way to its passage around the Sun and then, as it receded, far, far back into space.

So much of its path was carefully marked out that its entire orbit could be plotted with what seemed reasonable accuracy. That orbit turned out to be the most enormous ever calculated for any body in the solar system.

At its perihelion, Kohoutek's Comet, or Comet Kohoutek, as it is more commonly known, approaches to within 23 million miles of the Sun. This is less than half the closest approach that Halley's Comet makes to the Sun, and is even less than the closest approach of the planet Mercury, or of Encke's Comet. Still, Comet Kohoutek is no sun-grazer.

At the other end of Comet Kohoutek's orbit, however, when it is at aphelion and is farthest away, it is 334,400 million miles from the Sun. This is 102 times as far away from the Sun as Halley's Comet ever gets. It takes Comet Kohoutek over 200,000 years to go once about this mighty orbit.

Even Comet Kohoutek's aphelion need not be the limit. There may well be comets at even greater distances from the Sun. The nearest star, Alpha Centauri, is 4.3 light-years away. That is, it is at a distance that would take light 4.3 years to cross, and light travels at a speed of 186,282 miles per second. Alpha Centauri is thus about 25 million million miles away.

Any object that is even two light-years from the Sun (11.8 million miles) would remain in its gravitational grip. The Sun would be closer to it than Alpha Centauri would be at any time, and no other star would even come close to disputing the Sun's gravitational pull on such an object.

The study of Comet Kohoutek in 1973, shown in these University of Arizona photographs, helped astronomers to calculate the amount of time needed for this comet to complete one orbit around the sun—35,000 years!

The Daylight Comet of 1910 came in the same year as Halley's Comet. It is shown here, probably at dusk, in Algeria. The comet appeared in January, but was not as bright at northern latitudes as Halley's Comet of 1910.

An object two light-years away would be 37 times as far away as Comet Kohoutek at aphelion. Another way of putting it is that Comet Kohoutek recedes to a point only 1/18 of a light-year from the Sun.

The Estonian astronomer Ernst Julius Öpik (b. 1893) pointed all this out in 1930, at which time he speculated that there might be comets circling the Sun at these great distances. The idea was revived in 1950 by the Dutch astronomer Jan Hendrik Oort (b. 1900).

Oort pointed out that there may be very many comets circling the Sun at distances of from 1 to 2 light-years from the Sun. (Some people estimate that there may even be 100,000 million comets circling there. But even that many comets would have a total mass of not more than the planet Earth.)

Ordinarily, Oort suspects, those comets would be circling the Sun in orbits that are not very elliptical, so that they would stay very distant from the Sun at all times, and could not possibly be seen from the Earth. They would be little masses of icy substances from 1 to 10 miles across.

If it were not for outside influences, these distant comets would remain in their orbits for many billions of years, but there are the gravitational pulls of the nearer stars. These pulls are not strong enough to drag the comets away from the Sun, but they can be strong enough to change the orbits of some of the comets in minor ways, and such changes might accumulate with time. Some of the distant comets might be pulled forward, so as to move faster than before; some held back, so as to move more slowly than before.

Those that are made to move faster move away still farther from the Sun and may, in the end, be lost to the solar system forever. Those that are made to move slower will move closer to the Sun. If the slowing is sufficient, they may move close enough to the Sun to begin to be affected by the Sun's outermost planets. These might introduce further changes that would make the comet move quite close to the Sun at perihelion and even change its orbit into one that is never very far from the Sun even at aphelion.

The comets, in Oort's view, are distributed about the Sun in a huge spherical shell. This seems likely, since the new comets, those with very long elliptical orbits, can arrive from any part of the sky. The planets themselves all circle the Sun in very nearly the same plane. All the planetary orbits, in other words, would fit into a wide, shallow box in the shape of one in which pizzas are delivered.

The comets can move well outside such a box, however. They may travel at right angles to the general planetary plane, or even be moving in the opposite direction to the general planetary movement. Seen from far above Earth's north pole, all the planets move about the Sun in a counterclockwise fashion (opposite to the direction in which the hands of a clock move). Halley's Comet, however, moves clockwise, in the direction opposite to that of the planets.

If all the distant comets are distributed about a sphere, there is a chance that every once in a while two comets may collide as they revolve about the Sun. In the process, the two motions may partly cancel each other so that both may plunge into orbits far closer to the Sun.

Oort has estimated that in the time during which the solar system has so far existed, about one-fifth of the total number of comets has left the shell one way or the other. Some have escaped from the solar system and

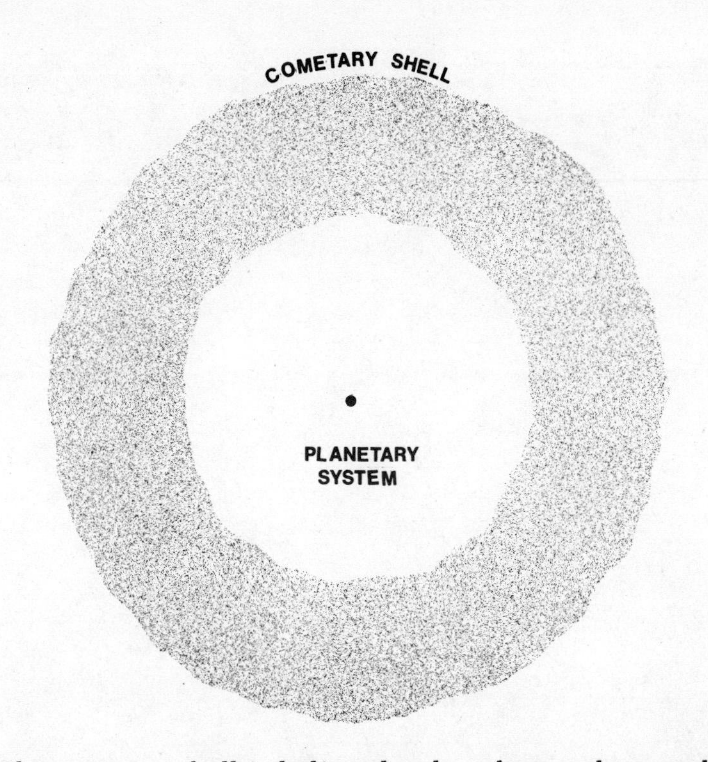

The cometary shell is believed to be a huge sphere at the outer reaches of the solar system. It consists of a thin cloud of billions of comets moving in orbits that are one to two light-years from the Sun and the planets, which would comfortably fit into the black dot at the center of this sphere! Occasionally, a comet will be dislodged from its remote orbit, swing in towards the Sun, and be visible from the Earth.

some have moved closer to the Sun, been affected by the planets, and put into a short-period orbit about the Sun, and would then have disintegrated in time. That, however, still leaves many, many billions in the shell to supply us with innumerable new visible comets in the Earth's sky.

In May 1910 Halley's Comet was seen over Paris in the early morning sky.

9

The Birth of Comets and the Solar System

Although no one has actually detected the comet shell about the Sun, there is enough evidence for one so that most astronomers seem quite certain that Oort is right. But if he is, where do all these comets come from?

The general feeling among astronomers now is that the entire solar system originated about 4,600 million years ago from a vast, slowly whirling cloud of dust and gas that gradually contracted under the pull of its own gravitation.

The central regions of the cloud contracted, whirling faster and growing denser and hotter as they did so, until, at the very center, the density and temperature became great enough for nuclear reactions to begin taking place. The nuclei of the hydrogen atoms that made up about 90 percent of all the atoms in the cloud smashed together energetically enough to fuse into helium atoms.

Such fusion releases considerable energy, and all the fusions together made the center of the cloud begin to glow. The cloud, in other words, had undergone "nuclear ignition" and had become the Sun.

On the outskirts of the condensing cloud of dust and gas, matter remained much less dense, of course, and much cooler. There was turbulence in this outer region of the cloud, and eddies of material were set up. Where neighboring eddies brushed against each other, matter in the two eddies collided and clung together so that larger bodies began to form there. These eventually became the planets, which thus came into existence at about the same time as the Sun did, but were never actually part of the Sun.

Meanwhile, on the very rim of the original spherical cloud of dust and gas was material that didn't share in the contracting process. It was too far away from the center to be sufficiently affected by gravitational pull, and it tended to remain where it was. Turbulence made it coalesce into small, icy bodies, and those are the comets that still remain in the vast shell about the Sun that marks the edges of the original cloud.

From this view, the distant comets can be of great interest to astronomers, who are curious as to the original makeup of the cloud of dust and gas that went into the formation of the solar system.

The Sun, of course, was originally a good sample of this material, but for 4,600 million years it has been undergoing hydrogen fusion and giving off a solar wind. We can't be sure, therefore, how closely it still resembles the makeup of the original cloud.

The inner bodies of the solar system are completely different in makeup from that of the original cloud. The original cloud, astronomers have reason to think, was about 90 percent hydrogen, 9 percent helium, and 1 percent everything else. The inner bodies of the solar system, however, were so warm from the glow of the

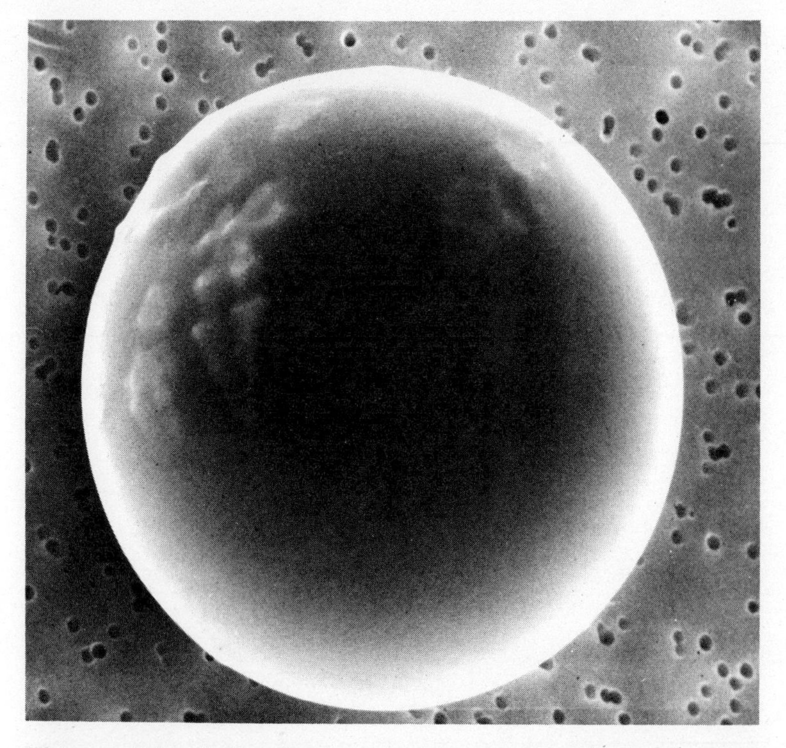

This particle of dust from a passing comet, collected by high altitude NASA aircraft and magnified 7,500 times, is approximately .0025 inches in diameter. It may hold chemical clues to the formation of the solar system.

nearby Sun that they tended to lose the light elements, hydrogen and helium. This loss was aggravated by the solar wind, which was particularly strong in the early days of the solar system, and which swept away the hydrogen and helium to regions beyond the asteroid belt: that is, the region where numerous small bodies, or asteroids, circle the Sun between the orbits of Mars and Jupiter.

As a result, the inner bodies of the solar system, from Mercury to the asteroids, are, for the most part, made up of the 1 percent of the substances other than hydrogen and helium. They give us very little clue to what the original state of the solar system may have been like.

The large planets of the outer solar system contain quantities of hydrogen and helium. Jupiter, in particular, seems to be made up of these substances almost entirely and to be representative of the original matter of the solar system. What's more, there have been no nuclear reactions taking place within Jupiter to change matters. However, Jupiter is a large body and there is reason to think that the matter within it has sorted itself out, with helium more concentrated toward the center and hydrogen more concentrated toward the surface, and other substances distributed we know not how. To get an idea of the original makeup of the solar system we'd need to know the makeup of every part of Jupiter, which would be hard to do.

The satellites of the outer planets have retained some of the light matter that originally existed in the cloud and have not lost much through the action of the solar wind—but in their early days, the giant planets must have had very high temperatures, and this would have affected the satellites to a greater or less degree. In addition, the satellites have been bombarded by small objects through the early period of their existence as they were gradually forming, and we don't know what changes may have been introduced in this way.

Of all the bodies in the solar system, the ones that are the least changed from the beginning are, undoubtedly, the distant comets. They are far removed from any source of heat or any significant disturbance by the solar wind, or from anything, in fact, but a very occasional

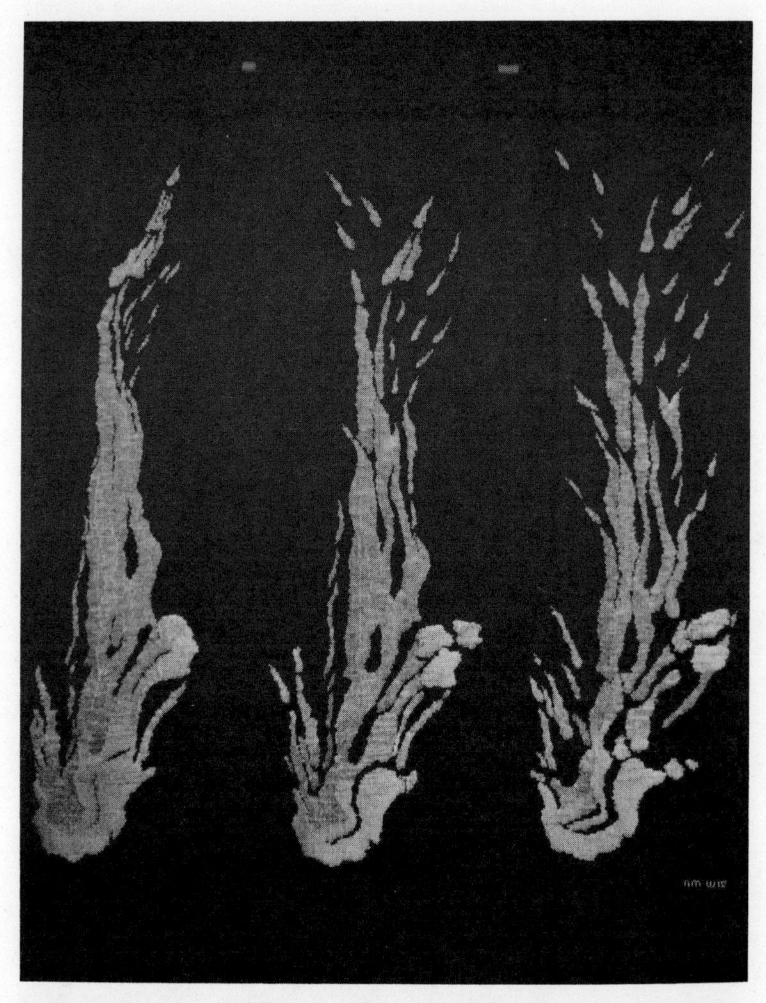

A contemporary California artist, April D. May, has created a modern tapestry—made of cotton, wool, and mohair—that depicts three views of a comet.

Comet Kohoutek gives off varying amounts of light at different points in its structure. The variations were color-coded and enhanced by computer in this 1973 photo taken from Skylab.

collision with one another. They have been in the deep freeze for 4,600 million years and are probably still exactly as they were in the beginning.

The trouble with that is we can't reach them, and it may be a long, long time (if ever) before we get to the point where we can reach them.

But they occasionally reach us. A comet that comes into the inner solar system brings its structure with it, giving us an opportunity to examine it. This has been done through the study of the spectra of comets. It would appear that comets are made up mostly of substances like water, ammonia, and methane. When they are heated by the Sun, fragments of molecules of these substances can be detected. In addition, traces of certain metals have been reported.

Comet Kohoutek was the first comet whose radiation of radio waves was studied. This made it possible to show the presence of hydrogen cyanide and methyl cyanide.

This is only the beginning. A comet may spend several weeks at distances such as are easily reached nowadays by rockets, and a variety of studies might be made at really close quarters. Not all comets, however, are equally suitable for the purpose.

Comets that have really small orbits and have visited the Sun many times are generally so altered as to be useless for study. Much of the icy material is vaporized and gone, and if anything is left at all it is only a rocky core.

On the other hand, new comets arising from the distant cometary shell, comets that have not yet had a chance to change much as a result of the Sun's heat, do not arrive at any foreseeable time. We have had no really good ones since 1882. Even Comet Kohoutek was a disappointment. Its brightness was far short of what had been predicted, judging from what seemed to be its size. Apparently it was rockier and less icy than the average comet would be so that it produced a smaller coma and a shorter tail.

What we need, ideally, is a comet that has not visited the Sun too often; yet is bright and comes on schedule, so that we have plenty of time to lay our plans and prepare for it. We need a short-period comet, but not an extremely short-period comet.

Of all such comets, Halley's Comet is the most nearly ideal. It is still bright. Though much of its icy material has gone and its structure cannot be exactly what it was in the days before it had passed the Sun closely, enough may be left to give us useful information.

More important, we know exactly when Halley's Comet may be expected.

As this book is being written, it is almost time for its return. Its next perihelion will be on (or very near) February 9, 1986. It will pass within about 57 million miles of Earth on November 27, 1985, on its way to perihelion, and within about 39 million miles of Earth on April 11, 1986, on its way back from perihelion.

This is not nearly as close as its approaches on occasions in the past. What's more, in April 1986, when Halley's Comet will be at its brightest, it will be very low in the sky in the hours just before dawn as viewed from Europe or the northern United States. In November 1985 it will be higher in the sky, but it will also be farther away. This means that people in the northern hemisphere are not going to have a very good show this time, and very few people now alive will have a chance to see it in 2063, when it next returns.

Thirty years ago, an American illustrator imagined a Buck Rogers culture on Earth watching with fascination as Halley's Comet returned in 1986.

Is it possible we may be cheated of even this much? Can an accident have happened to Halley's Comet since its last arrival? Apparently not. We have already seen it. It is on its way.

On October 20, 1982, astronomers managed to detect it by an electronic camera attached to the 200-inch telescope on Mt. Palomar. Its magnitude was 24.2, which meant it was only about a fifteen-millionth as bright as the faintest star that can be seen with the unaided eye. At the time of the sighting, it was 1,000 million miles from the Sun, somewhat farther away than the planet Saturn.

Can we be sure that what has been seen is Halley's Comet? Yes, we can. It was located in almost exactly the spot where it was supposed to be, and it has been growing brighter ever since and has been following the proper orbit. It's Halley's Comet, all right.

The European Space Agency is scheduled to send out a flyby spacecraft, or "probe," named "Giotto," to study Halley's Comet, and the Soviet Union and Japan are each also making plans to send out probes. The United States, however, scrapped its plans in this direction, at least as far as a direct probe is concerned. However, it may be possible to divert a planned Venus Pioneer Orbiter probe and have it pass Halley's Comet en route to Venus. In addition, an earlier Explorer rocket will make observations of a dim comet that will swing around the Sun a few months before Halley's Comet does.

However poor the spectacle of Halley's Comet may be this time from Earth's surface (especially in the northern hemisphere), the spacecraft will be able to take photos that should show a comet in more spectacular fashion than one has ever been seen before. The spectra can be taken with greater accuracy, so that not only will we learn much more about the composition of the ices, but also something about the rocky material embedded in the ice and floating in the coma.

In addition the nucleus will be examined, as will the nature of the tail, so that studies can be made of changes in the cometary structure as it approaches the Sun. There will also be studies of its rotation, together with all sorts of things that may turn up unexpectedly and cannot be predicted (as was true, for instance, of the craters on Mars, the volcanoes on Io, and the structure of the rings around Saturn).

10

Comets and Catastrophes

We are now all set to reach out and virtually touch a comet; we are getting prepared to learn more about comets in one rocket trip than we have learned in all the previous centuries of civilization. And yet, now that all this is true, there has come a return of the old, old comet fear.

It is not the ancient kind, for it is not a superstitious consideration of omens. It isn't even the old biblical kind, for we know that comets are small objects. There isn't a chance that any comet's tidal effects would produce a flood, or that its tail colliding with the atmosphere would produce rain, as William Whiston suggested nearly three centuries ago. The thought of a comet pulling us into the Sun, as he predicted one would, is even more laughable.

No, what now concerns people is the possibility of collision. The chance of collision with any one comet is virtually zero, but if there are a great many comets, the chance of collision with one of them, sooner or later, is

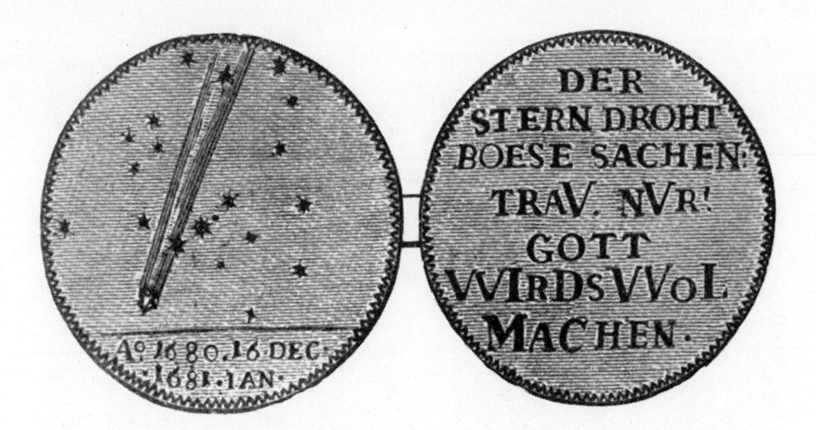

These coins, distributed by monks in 1680, read: "This star threatens evil things. Trust in God, who will turn them to good." The Roman capitals in the German script add up to 1681, the year of the Comet's appearance. The medals were designed to ward off the evil influences warned of by the astrologic beliefs of the seventeenth century.

very great. If there are a hundred billion of them out there and if several billion of these have already entered the planetary system during the history of the solar system, surely collisions must already have taken place in the past.

There are indeed signs of craters on the Earth's surface, most of them so nearly obliterated by the action of wind, water, and life that they are almost unnoticeable except from the air, and then only because of the presence of circular lakes and of other giveaway signs. One clearcut example is Meteor Crater in Arizona, which has remained largely uneroded because it was formed only a few thousand years ago in an area where there is little water and little life.

Meteor Crater in Arizona, photographed here from 2000 feet, was formed a few thousand years ago by the impact of a small asteroid, not a comet, on the Earth's surface.

These craters are formed by large meteors (or, if you prefer, small asteroids).

Comets, of course, are made up of icy matter that is neither as dense or as tough as are the rocks and metals making up meteors, so that a collision with a comet may not seem as fearful as one with a meteor. That, however, is not necessarily so.

Consider the case, for instance, of an event that took place on June 30, 1908. On that day a fireball lit the sky in broad daylight over a desolate forested region of eastern Siberia near a place called Tunguska. There was a tremendous explosion and every tree within twenty miles was killed and knocked down. A herd of 1500 reindeer was wiped out. A man who was 50 miles away was knocked down. Fortunately, there were no human beings nearer than that and not one person was killed. If an explosion of that size had taken place in a large city, however, millions of people would have been killed instantly.

The place where this happened was so inaccessible that no investigators were able to get there, especially since World War I began soon afterward, followed by the Russian Revolution and a civil war. It was decades before anyone finally struggled to the site.

Everyone would assume that a large meteorite had hit, but no meteorites of any kind were found at the site. There wasn't even a crater. Apparently whatever had exploded had done so in midair.

But what could explode in the air and leave no sign of itself? Suppose the Earth had collided with a small comet. It would pass through the air rapidly and be heated to the point where all its icy substance would vaporize at once in a vast explosion. There would be nothing left behind but vapor, which would mix with the atmosphere.

It must have been a small comet. Astronomers estimated that to do the damage it did, it would have to have been about 200 feet across. Most comets seem to be considerably larger than that, so the Tunguska explosion may have been caused by a mere piece of a comet, a piece that had been broken off a much larger comet because of solar heating at perihelion. The piece broken off would have moved along with the parent body but might slowly have been driven farther off by the rocket effect of evaporation into an unfortunate collision course with Earth. It is possible, by reconstructing its orbit, that the comet fragment had been part of Encke's Comet to begin with.

If a comet fragment could do such damage, what would happen if a full-sized comet struck?

In that connection, we must consider an event that took place about 65 million years ago. At that time the most remarkable animals on Earth were the dinosaurs. These were reptiles, and they included some of the largest and most magnificent of all animals who had ever lived upon dry land.

There had been a great many kinds of dinosaurs. Some kinds became extinct and were replaced by others; but for 100 million years many kinds dominated the land. Other large reptiles lived in the sea, and some even flew through the air.

Yet 65 million years ago, something happened. All the dinosaurs died out over a very short period of time. So did the great reptiles of the sea and air. So did many other kinds of animals and even many kinds of microscopic creatures. It was a "Great Dying." Some biologists estimate that 75 percent of all the different kinds of living things that dwelt upon the Earth were wiped out, and that the remaining 25 percent that managed to survive did so narrowly.

Scientists did not have any idea what might have caused this Great Dying. Many theories were advanced, but none were completely satisfactory.

In 1979, however, an American scientist, Walter Alvarez, was working with ancient rocks in central Italy and found a thin section where the rare metal iridium was 25 times as high as it was below or above in the

This devastation of a forest in Tunguska, Eastern Siberia, in 1908, is believed to have been caused by the tremendous explosion of a comet fragment that vaporized seconds before it would have hit the Earth.

rock. This was sedimentary rock, consisting of mud that had slowly settled out and been buried and compressed under other layers. But why should there have been a brief period when some mud that had settled out be unusually high in iridium?

There are good techniques for measuring the age of that section of the rock, and it turned out to be 65 million years old. It had been laid down just at the time of

There is clear evidence that 65 million years ago the dinosaurs and perhaps as much as 75 percent of all species were wiped out. Was this "Great Dying" due to a surge of comets that bombarded the planets?

the Great Dying that had killed off all the dinosaurs. Surely that could not be mere coincidence. There had to be a connection.

Alvarez suggested that a meteorite must have struck the Earth at that time. Meteorites are richer in iridium than the Earth's crust is, since on Earth most of the iridium mixed with iron and settled to the Earth's center, where there is a nickel-iron core.

The meteorite must have been large enough to vaporize cubic miles of ground where it struck. The vapor exploded high into the stratosphere, cooled into dust, spread around the world, and then settled down everywhere, carrying with it its load of iridium.

Upon further investigation, this theory looked better and better, for when people dug up rocks in other places in the world, they often found an iridium-rich layer just at the 65-million-year mark. What's more, other metals were increased in concentration there too, and were present in much the same proportion as they are in meteorites.

But why should that have destroyed the dinosaurs? The meteorite would have formed a large crater and destroyed everything for miles around, but why should the whole Earth be affected? Alvarez suggested that the dust that spread through the stratosphere blocked the sunlight for a considerable period of time, so that there was a long darkness. Most plant life died without sunlight, and then the animal life that lived on plants also died, and then the animal life that lived on animals.

Some plants lived on in the form of seeds and roots and began to grow after the dust finally settled out and the Sun shone again. Some animal life survived too, especially small animals who managed to live on by eating remnants of dead or dying plants, or the carcasses of the many dead animals that were frozen and preserved in the cold that had settled down over the Earth with the disappearance of sunlight. When the sunlight returned, the Earth was soon repopulated, but with types of animals different from the old. Mammals and birds began to dominate the land instead of reptiles.

Once scientists began to think about this possibility, the question arose as to other Great Dyings that had taken place. The one that killed the dinosaurs was the most dramatic and the best-known, but it wasn't the only one.

In fact, as the evolutionary record was studied very carefully, it seemed that there were Great Dyings every 26 million years or so.

But why should there be such regular Great Dyings? We don't know of anything else here on Earth—or out in space, for that matter—that takes place every 26 million years, and is deadly for life on Earth. If the dinosaur Great Dying was caused by the collision with Earth of some object from space, why should such collisions take place every 26 million years, unless something in space has a period that long?

In 1983, an interesting suggestion was advanced by several American scientists.

Suppose that there was a large planet or a tiny, very dim companion star that circled the Sun at such an enormous distance that we couldn't possibly become aware of its existence without a very careful search of a kind that has never yet been made.

And suppose that the companion was so far away that it had a very long elliptical orbit, rather like that of a comet. Every 26 million years it might approach the

This 1858 drawing by the celebrated French caricaturist Daumier bears the caption: "Ah, comets . . . that always means bad luck! No wonder poor Madame Galuchet died suddenly last night!"

Sun, and then it might have some effect on the solar system that would result in Earth being struck by objects from space.

Suppose, for instance, that at perihelion this companion object passed through the asteroid belt. It would then disrupt the orbits of many of the hundred thousand or so asteroids and send some of them further toward the Sun. One of them might hit the Earth and might cause a Great Dying.

However, a dwarf star passing through the asteroid belt would also modify the orbits of the planets, including that of Earth, and there is no sign that this has ever happened.

But suppose that the companion's orbit carried it out to a distance of about 2 light-years from the Sun at aphelion and to just under 1 light-year at perihelion. It would take about 26 million years for an object to go once about such a mighty orbit.

At perihelion, the companion would pass through the innermost portion of the comet shell, where the comets were thickest.

Millions of comets might then have their orbits disturbed and would move down among the planets. Bright comets would appear in the sky in rapid succession, week after week, for a thousand years or more, and a few would be quite likely to hit Earth during that period, with disastrous results, The hits have not wiped out life altogether, but on several occasions they nearly succeeded in doing so.

When the comets start appearing in the sky so frequently, they can be looked upon as omens of disaster indeed.

It is no wonder that the name "Nemesis," after the Greek goddess of vengeful destruction, was suggested for the companion star by the American scientist Richard A. Muller. There is no direct evidence so far that it really exists, of course, but we may expect astronomers to look for it now, with all the new tools at their disposal.

Even if Nemesis exists, there is no need for immediate fright. The last Great Dying was 11 million years ago, so Nemesis must be near its aphelion now (and must be

Comets have long been linked with a darkness of the soul. In this renowned print by Albrecht Dürer, the Dutch artist, melancholy is symbolized by a blazing comet.

particularly hard to spot). The next Great Dying would not come for some 15 million years or so.

By that time, if the human race has not destroyed itself, by nuclear war or otherwise, we should be sufficiently advanced technologically to maintain a space watch on all cometary orbits. If we should detect one that shows signs of being dangerous for Earth, it might then be the task of the spaceships on watch to destroy that comet with a nuclear bomb, or something more advanced, thus converting it into a cloud of rocky grit that would do no more to Earth than give it an amazing shower of shooting stars.

Indeed, we might even argue that this new suggestion as to the cause of the Great Dyings has already contributed to saving humanity. The description of the "comet winter," in which the dust thrown up into the stratosphere brought a long, cold night to Earth, has encouraged such scientists as the American Carl Sagan (b. 1935) to consider just exactly what might happen if nuclear bombs exploded. (He was aided here by the studies he had made of planetary dust storms on Mars, which our rocket probes had carefully observed.)

He concluded that in a nuclear war, hydrogen bombs would send enough dust into the stratosphere to create a "nuclear winter" that would produce a Great Dying of its own. In a nuclear war all would lose, those who fought the war and those who merely watched.

The possibility of such a nuclear winter is further terrifying many human beings who were already frightened of nuclear war. If this new fear hardens public opinion and forces the superpowers into an agreement that would make nuclear war impossible, then the fear of comets that has so long plagued humanity would, in the end, save us all.

List of Illustrations

Picture Credit Key

AMNH: Courtesy of the Library Services Department, American Museum of Natural History.
DM: Kenneth Heuer, *Wonders of the Heavens*. New York: Dodd, Mead, & Company, 1956. Permission granted by the publisher.
HC: T. E. R. Phillips and W. H. Steavenson, eds. *Splendour of the Heavens*. London: Hutchinson and Company, 1923.
HP: Permission granted by the Richard S. Perkin Library, American Museum of Natural History, Hayden Planetarium, New York.
NASA: National Aeronautics and Space Administration, Washington, D.C.
NS: American Numismatic Society, New York.
OUP: George F. Chambers, *Story of the Comets*. London: Oxford University Press, 1910. Permission granted by the publisher.
SK: Sandra Kitt, illustrator.
YO: Yerkes Observatory, Williams Bay, Wisconsin.

ii Halley's Comet, 1910, Yerkes Observatory: AMNH
x Mallet drawing in *Description de L'Univers*. Paris, 1683: HP
2 Title page from George Henischius, *Description of the Comet of 1596*. Augsburg: Johann Schultes: HP
3 Ambroise Pare, *Livres de Chirurgie*. Paris, 1597: OUP
5 "Forms of Comets" from Johannes Hevelius, *Cometographie*. Danzig: Simon Reiniger, 1668: HP
6 Halley's Comet, A.D. 66, in Camille Flammarion, *L'Astronomie*. Paris, 1880: HC
9 Path of Halley's Comet through the sky in 1985–1986, a diagram: SK

10 Woodcut of Halley's Comet, August 1531: NASA
13 Great Comet of 1577 in an article by Henry Norris Russell, "The Heavens in July," in *Comets*, edited by John C. Brandt. New York: *Scientific American, Inc.*, 1910. All rights reserved.
14 Elliptical and parabolic paths, a diagram: SK
17 Path of Halley's Comet through solar system—1984–2004, a diagram: SK
18 Halley's Comet, 1531, from Camille Flammarion, *L'Astronomie*, Paris, 1880: HC
19 Facsimile from page 900 of Gregory's *Astronomy*, 1726, printed with permission from the *Journal of the Royal Astronomical Society of Canada*.
20 Portrait of Edmund Halley: OUP
21 Halley's Comet, 1456, from S. De Lubieniecki, *Theatrum Cometicum*, Amsterdam, 1668: HC
27 "The Adoration of the Magi," 1302–1304, by Giotto: HP
28 Halley's Comet, 1066, from the Bayeux Tapestry: OUP
29 Halley's Comet in 1066, from S. De Lubieniecki's *Theatrum Cometicum*, Amsterdam, 1668: HC
31 Halley's Comet, 684, in a woodcut by Michel Wohlgemuth from the *Nuremberg Chronicles*, 1493: HC
32 Dutch coin from 1577: NS
33 German coin from 1910: NS
34 "The Ides of March" painted by Sir E. J. Poynter: HC
39 From *Le Ciel et L'Univers* by Théophile L'Abbe Moreux: AMNH
40 *Punch*, December 5, 1905: OUP
41 Morehouse's Comet, 1908, taken at Yerkes Observatory: AMNH
42 Plate showing Chinese during the Great Comet of 1910, from *Romance of Modern Astronomy*, London: Seeley, Service, and Company, 1913: HP
43 Halley's Comet, 1910, taken at Yerkes Observatory: AMNH
47 Drawings of Halley's Comet, 1835–1836, by C. P. Smyth: OUP
48 Great Comet in 1861 drawn by Warren De La Rue: HP
51 Halley's Comet, 1910: YO
53 Halley's Comet, 1910, Mt. Wilson/Palomar Observatory: AMNH
54 Great Comet of 1811, from R. A. Proctor, *Flowers of the Sky*. London: Sampson, Low, Marston, Searle, Rivington, no date: HP

57 Comet of 1843 over Paris, from Amedée Guillemin, *The World of Comets*. New York: A.C. Armstrong and Son, 1877: HP
58 Donati's Comet of 1858, from Dr. Edmund Weik, *Bilderatlas der Sternenwelt*. Stuttgart: J.F. Schreiber, 1888: HP
61 Great Comet of 1861, from Dr. Edmund Weik, *Bilderatlas der Sternenwelt*. Stuttgart: J. F. Schreiber, 1888: HP
62 Solar eclipse: OUP
65 Tails away from sun, from Camille Flammarion, *L'Astronomie*. Paris, 1880: OUP
68 Great Comet of 1744, drawn by Cheseaux: HP
69 Cheseaux's Comet at Lausanne, Switzerland: OUP
72-3 Fourteen views of Halley's Comet, 1910, Mt. Wilson/Palomar Observatory: AMNH
74 Comet Mrkos, 1957, photographed from Mt. Wilson/Palomar Observatory: AMNH
75 Meteor showers, a diagram: SK
77 Meteor shower, from Asa Smith, *Smith's Illustrated Astronomy*. Boston: Chase, Nichols, and Hill, 1948: AMNH
81 Three photographs of Comet Kohoutek: NASA
82 The Daylight Comet of 1910 drawn by W. B. Gibbs, F.R.A.S., from *Journal of the British Astronomical Association* 20 (April 1910).
85 Cometary shell, a diagram: SK
86 Halley's Comet, 1910, from Camille Flammarion, *L'Astronomie*, Paris, 1880: HC
89 Microphotograph, 1979: NASA
91 Weaving, 1982, by April D. May, photograph by Frank Talbott.
92 Comet Kohoutek, Skylab 1973: NASA
95 Return of Halley's Comet in 1986 drawing by Matthew Kalmenoff: DM
99 Medals from 1680, from *Popular Astronomy*, vol. 18 (1910). Permission by Goodsell Observatory, publisher.
100 Meteor Crater in Winslow, Arizona, photograph by Fairchild Aerial Survey, Inc.: AMNH
103 Tunguska, Siberia, 1908: AMNH
104 Dinosaur watching comet: AMNH
107 Daumier cartoon: HP
109 Dürer, "Melancholia I," 1514: HP

Index

Adams, John Couch, 76
Adoration of the Magi, The, 26, 27
Agrippa, Marcus, 4, 32
Alvarez, Walter, 102, 105
American Civil War, 59, 60, 61
Ammonia, 93
Andromedids, 76
Aphelion, 24
Apian, Peter, 8, 10, 25
Aquarids, 75, 76
Aristotle, 11, 12, 69
Astrology, 2, 3
Attila, 30, 31

Barnard, Edward Emerson, 60
Bayeux Tapestry, 28, 29
Bessel, Friedrich Wilhelm, 45
Biela, Wilhelm von, 45
Biela's Comet, 45–49, 74, 76
Bielids, 76
Biot, Jean Baptiste, 70
Borelli, Giovanni Alfonso, 15
Buffon, Georges L. L. de, 37

Caesar, Julius, 4, 34, 35
Calixtus III, 25
Calpurnia, 34
Chalons, Battle of, 31
Cheseaux, Jean de, 68
China, comet observation in, 30–35
Chladni, Ernst F. F., 70
Clairault, Alexis Claude, 23, 24
Coma, 50
Comet Encke, 42
Comet Halley, 24
Comet Kohoutek, 80, 83, 92, 93, 94
Comet Mrkos, 74
Comet of *44* B.C., 4, 35
Comet of *66*, 6, 32
Comet of *684*, 30, 31
Comet of *1066*, 28, 29
Comet of *1301*, 26, 27
Comet of *1456*, 8, 19, 21, 25
Comet of *1531*, 8, 10, 19, 25
Comet of *1577*, 11, 13, 32, 35
Comet of *1607*, 13, 18, 25
Comet of *1682*, 18, 21, 25
Comet of *1744*, 68, 69
Comet of *1811*, 54–56
Comet of *1843*, 56, 57, 60
Comet of *1861*, 48, 49, 59, 61, 63
Comet of *1882*, 60, 62
Comet of *1910*, 82
Comet pills, 63
Comets, 1
  birth of, 87ff.
  collisions with, 37, 39, 98ff.
  comas of, 50
  death of, 45ff.
  discovery of, 61
  distance of, 11, 12, 79–83
  distant shell of, 83, 85
  fear of, 1–7, 36, 37, 46, 98, 99, 107

Great Dyings and, 107, 108
long-period, 43, 44
meteor showers and, 74–78
new, 55
nuclei of, 50
number of, 7, 79
orbits of, 39ff.
paths of, 8ff., 13ff.
photographs of, 60–63
planetary pull on, 23, 24
radio waves and, 93
rocky fragments of, 67, 89
shape of, 4, 5
short-period, 43, 44
spectrum of, 59
structure of, 49, 50, 57, 58, 90, 93
tails of, 5, 8, 10, 50–52, 64–69
telescope and, 15
Comet wine, 55
Constantinople, 7, 25
Copernicus, Nicolas, 12
Cysat, Johann, 15

Daumier, Honore, 107
Dinosaurs, 102, 104
Donati, Giovanni Battista, 56, 58
Donati's Comet, 56–59
Durer, Albrecht, 109

Ellipses, 13–16
Encke, Johann Franz, 41
Encke's Comet, 40–42, 52
Tunguska event and, 102
England, 6, 7, 28

Flood, Noah's, 37
Foci, 13
Fracastoro, Girolamo, 8, 25

Galileo Galilei, 14
Gauss, Karl Friedrich, 41
Gill, David, 60
Giotto, 96
Giotto de Bondone, 26, 27
Gravitation, universal, 16
Great Dyings, 102ff.
periodic, 106
Guericke, Otto von, 16

Halley, Edmund, 18–22, 25, 33
Halley's Comet, 6, 24, 52, 53, 62, 63
aphelion of, 25
brightness variation of, 26
commemorative coin for, 33
Constantinople's fall and, 7, 25
drawing of, 30, 31
England and, 6, 7, 28
fear of, 42, 63
Huns and, 31
in *1758*, 23
in *1832*, 46, 47
in *1910*, 62, 63
in *1986*, 94–97
Jerusalem and, 32
Mark Twain and, 47, 63
meteor showers and, 75, 76
Nativity and, 26, 27, 32, 35
Noah's Flood and, 37
orbit of, 23
path across the sky of, 9, 19
path around the sun of, 16, 17
perihelion of, 23, 24
period of, 26
photographs of, 43, 47, 51, 53, 72, 73
pronunciation of, 25
return of, 22ff.
sighting in 1982 of, 96
space probes and, 96, 97
various appearances of 25–35
Harold, 28
Huggins, William, 59
Hydrogen, cyanide, 93

Iridium, 102, 103, 105

Jerusalem, 6
Josephus, 32
Jupiter, 90
  comets and, 40

Kepler, Johannes, 12–15, 25, 79
Kirkwood, Daniel, 74
Kohoutek, Lubos, 79
Kohoutek's Comet, 80

Lalande, Joseph Jerome, 23, 24
Lebedev, Peter Nicolaevich, 66
Leonids, 71, 76, 77
Leverrier, Urbain J. J., 74
Lexell, Anders Johan, 39
Lexell's Comet, 39–41, 46
Long-period comets, 43, 44
Ludwig the Pious, 4

Mars, 110
Maury, Matthew Fontaine, 47, 49
Maxwell, James Clerk, 66
Messier, Charles, 24, 38, 45
Meteor Crater, 99, 100
Meteorites, 70
Meteors, 69ff.
Meteor showers, 70ff.
Methane, 93
Methyl cyanide, 93
Milne, Edward Arthur, 66
Moon, 1, 14
  parallax of, 12
Morehouse's Comet, 41
Muller, Johann, 8

Napoleon I, 56
New comets, 55
Newton, Isaac, 16, 18
Noah's Flood, 37
Normans, 7
Nuclear ignition, 87
Nuclear winter, 110
Nuclei, comet, 50

Olmsted, Denison, 71
Oort, Jan Hendrik, 83, 85
Opik, Ernst Julius, 83
Orbits, 12
Orionids, 75

Palitzsch, Johann Georg, 23
Parabola, 14–16
Parallax, 11, 12
Perihelion, 23
Perseids, 74
Planets, 1
  birth of, 88
  orbits of, 12, 13
  structure of, 88–90
Pons, Jean Louis, 41
Portugal, 55
Protestant Reformation, 7

Radiation pressure, 66
Radio waves, 93
Regiomontanus, 8, 25
Rossi, Bruno, 67

Sagan, Carl, 110
Scheuchzer, Johann Jakob, 70
Secchi, Pietro Angelo, 49
Shiaparelli, Giovanni Virginio, 74
Shooting stars, 69
Short-period comets, 43, 44
Solar system, 12, 87ff.
Solar wind, 66, 67
Space probes, 96, 97
Spectra, 59
Star of Bethlehem, 26, 27
Stars, 1
Sun, 1, 4
  birth of, 87
  comets' tails and, 64–69
  possible companion of, 106–108

solar system and, 12ff.
structure of, 88, 89
Sun-grazers, 56, 57, 60

Tails, comets', 5, 8, 10, 50–52, 64–69
Telescope, 14, 15
Tunguska event, 100–103
Turks, 7, 25
Tycho Brahe, 11, 12

Venus Pioneer Orbiter, 96
Vienna, 7

Water, 93
Weiss, Edmund, 76
Whipple, Fred Lawrence, 49, 50, 59
Whiston, William, 36, 37, 98
William of Normandy, 6, 28